Management in the Built Environment

Series editor

Low Sui Pheng, National University of Singapore, Singapore, Singapore

The aim of this book series is to provide a platform to build and consolidate a rigorous and significant repository of academic, practice and research publications that contribute to further knowledge relating to management in the built environment. Its objectives are to:

(1) Disseminate new and contemporary knowledge relating to research and practice in the built environment
(2) Promote synergy across different research and practice domains in the built environment and
(3) Advance cutting-edge research and best practice in the built environment

The scope of this book series is not limited to "management" issues per se because this then begs the question of what exactly are we managing in the built environment. While the primary focus is on management issues in the building and construction industry, its scope has been extended upstream to the design management phase and downstream to the post-occupancy facilities management phase. Management in the built environment also involves other closely allied disciplines in the areas of economics, environment, legal and technology.Hence, the starting point of this book series lieswith project management, extends into construction and ends with facilities management. In between this spectrum, there are also other management-related issues that are allied with or relevant to the built environment. These can include, for example cost management, disaster management, contract management and management of technology.

This book series serves to engage and encourage the generation of new knowledge in these areas and to offer a publishing platform within which different strands of management in the built environment can be positioned to promote synergistic collaboration at their interfaces. This book series also provides a platform for other authors to benchmark their thoughts to identify innovative ideas that they can further build on to further advance cutting-edge research and best practice in the built environment.

If you are interested in submitting a proposal for this series, please kindly contactthe Series Editor or the Publishing Editor at Springer:

Sui Pheng Low (bdglowsp@nus.edu.sg) or
Ramesh Premnath (Ramesh.premnath@springer.com)

More information about this series at http://www.springer.com/series/15765

Low Sui Pheng · Benjamin K. Q. Chua

Work–Life Balance in Construction

Millennials in Singapore and South Korea

Springer

Low Sui Pheng
Department of Building
National University of Singapore
Singapore, Singapore

Benjamin K. Q. Chua
Department of Building
National University of Singapore
Singapore, Singapore

ISSN 2522-0047 ISSN 2522-0055 (electronic)
Management in the Built Environment
ISBN 978-981-13-1917-4 ISBN 978-981-13-1918-1 (eBook)
https://doi.org/10.1007/978-981-13-1918-1

Library of Congress Control Number: 2018950803

This Springer imprint is published by the registered company Springer Nature Singapore Pte Ltd.
The registered company address is: 152 Beach Road, #21-01/04 Gateway East, Singapore 189721, Singapore

Contents

Abbreviations

BSD	Boundary Spanning Demand
BSR	Boundary Spanning Resources
CI	Construction Industry
CS	Coping Strategies
GDP	Gross Domestic Product
LD	Life Demand
LR	Life Resources
LTO	Long-Term Orientation
MOM	Ministry of Manpower
OECD	Organization for Economic Co-operation and Development
PM	Proposed Measures
UA	Uncertainty Avoidance
WD	Work Demand
WLB	Work–Life Balance
WLC	Work–Life Conflict
WLE	Work–Life Enrichment
WLI	Work–Life Interface
WR	Work Resources

List of Figures

List of Tables

Chapter 1
Introduction

Abstract This chapter explains the structure of the book and introduces the background of this study as well as sets out the research problem, aim, objectives and scope of the study.

1.1 Background

The ability to achieve Work–Life Balance (WLB) has become an important topic in many societies in the world due to the societal and technological changes over the last few decades. The WLB concept has generated substantial interest in the academia, research and published literature. It has been said that achieving WLB results in positive outcome of an individual including job satisfaction, life satisfaction, organizational commitment, and psychological well-being. WLB research began gaining popularity in recent years despite its long history in the existing literature and research. One of the main reasons for this phenomenon is due to the demographic changes as well as the shifts in attitudes which are changing the workforce's perspective in the need and desire for WLB (Lingard & Francis, 2009). It has also been said that work and life imbalance has a detrimental effect to one's physical and psychological health (Lingard & Francis, 2009). The shift in this trend is a worrying issue especially in the construction industry where it has often been labelled by many people as dirty, demanding and dangerous (Ofori, 2000). The construction industry has also been characterized for its long working hours, high demand and stressful job scope (Lingard & Francis, 2009). Juggling work and life for employees working in the construction industry (especially in main contractor firms) is frequently tough and difficult. According to Bannon, Ford, and Meltzer (2011), WLB is one of the major concerns for the Millennials workforce. The Millennials workforce tends to avoid working long hours at the expense of spending time with their family and friends. With more Millennials employees entering the workforce, concerns related to WLB must be addressed by the organizations to help this generation achieve WLB.

L. Sui Pheng and B. K. Q. Chua, *Work–Life Balance in Construction*,
Management in the Built Environment,
https://doi.org/10.1007/978-981-13-1918-1_1

Even though Singapore has enacted laws to protect the workers from excessive working hours under the Employment Act, the nature of the construction industry has often pushed workers to the limit. The Ministry of Manpower (2011) in Singapore has also reported that construction industry practices have made employees work for weeks on end without any rest days. Working long hours without rest will also affect their physical and mental health. Most organizations (including building contractors) have failed to implement proper policies and practices that would help the workers eliminate or overcome this issue. These negativities have resulted in deterring the current Millennials from entering to work in the construction industry based on the figures reported by the Ministry of Manpower (2016) in Singapore. With an ageing workforce on the rise and the disproportional local to foreign workforce currently in the Singapore construction industry, it is clear that the relevant authorities and organizations should set up new policies that will help promote WLB especially to the Millennials building professionals.

1.2 Research Problem

WLB as mentioned earlier has generated substantial interest in the academia, research and published literature. Unfortunately till date, WLB still appears to be an underdeveloped and fragmented concept that has not been keeping pace with the increasing interests generated (Grzywacz & Carlson, 2007). Furthermore, there is still a lack of consensus on how WLB should be defined and measured despite the abundance of WLB studies in the existing literature (Rantanen, Kinnunen, Mauno, & Tillemann, 2010). Currently there is still no strong theoretical understanding of WLB, resulting in literatures and notions driven by diverse empirical findings. Likewise, the absence of strong empirical support has also resulted in the loosely defined conceptualization of WLB which came under the scrutiny of many researchers (Carlson, Grzywacz, & Zivnuska, 2009; Grzywacz & Carlson, 2007). The lack of empirical evidence can impair the ability of human resource practitioners to apply effective and viable organizational strategies to promoting WLB; thus causing counterproductive results. In addition, many studies generalized WLB as a single solution which can be applied to all individuals across different societies (Grzywacz & Carlson, 2007; Rantanen et al., 2010). However, this may not be feasible as there are also other elements that can influence the realization of WLB. Hence, this warrants an investigation to evaluate if achieving WLB can indeed be of a universal application or an isolated solution based on a case-by-case basis.

1.3 Research Aim and Objectives

The research aims to operationalize the concept of WLB which can be used to apply across different areas of study. A comparative study will be performed to validate the operationalized concept as well as to justify whether achieving WLB is indeed a universal application or an isolated solution to be considered on a case-by-case basis.

The research objectives presented in this study are as follows:

- To develop a framework that provides a linking mechanism to explaining WLB; and
- To evaluate and assess WLB as perceived from two different countries (namely Singapore and South Korea) with the same demographics in order to establish a fair comparison and interpretation of the results during the assessment

1.4 Research Hypotheses

The conceptual framework (presented later in the book) formed the basis for the research hypotheses. There are a total of 12 hypotheses in this research which will be reviewed and presented in the book.

1.5 Research Scope

This research focuses on operationalizing the concept of WLB and to test the conceptual framework by conducting empirical studies in the construction industry. However, the scope of this study is limited to assessing the level of WLB on Millennials employees working in the construction industry, particularly with the main contractor firms dealing with builder's works. In addition, the study will be targeting Singapore and South Korea nationalities, as well as to examine whether the application of the results on the conceptual framework of WLB will yield different outcomes in two different countries.

1.6 Structure of Book

This book comprises of seven chapters. Figure 1.1 illustrates the structure of the book. This chapter introduces the background of this research as well as setting the research problem, aim, objectives and scope of the study. Chapters 2–4 present the literature review of this study. Chapter 2 on Work–Life Balance and Work–Life

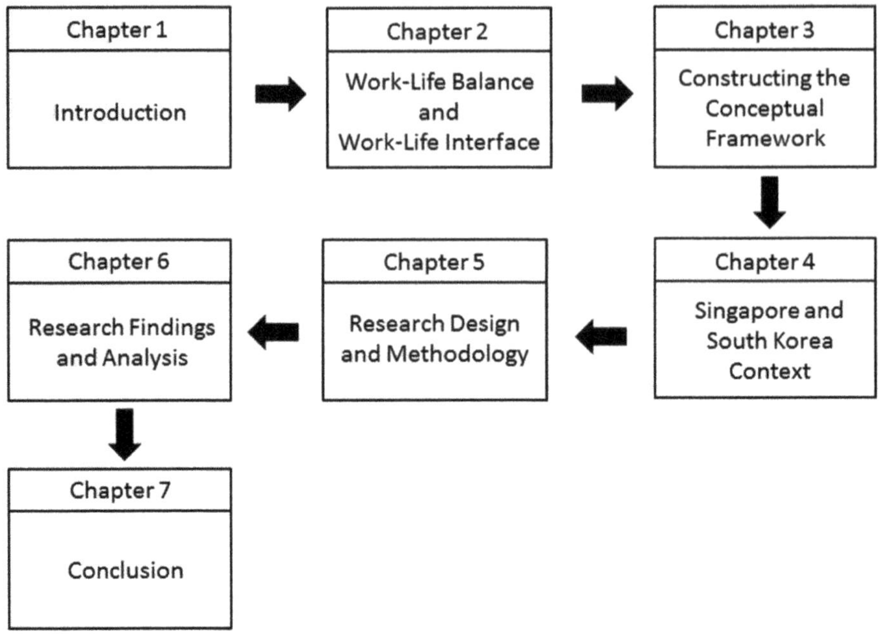

Fig. 1.1 Structure of the book

Interface provides insights on past WLB research works carried out as well as discussing all the Work–Life Interfaces that are associated with WLB. Chapter 2 covers the literature review associated with Work–Life Balance as well as developing the background knowledge of the WLB research which is essential for developing the conceptual framework in Chap 3. In Chap. 3, Constructing the Conceptual Framework, a conceptual framework is formulated based on the ideas and theories drawn from past research studies. This chapter also explains each of the indicators used in the conceptual framework in details. The research hypotheses are also presented to show the relationships between the succeeding and preceding attributes. Chapter 4 on the Singapore and South Korea Context introduces the background of the research scope and comprises of three sections: namely overview of Singapore and South Korea society, the construction industry in Singapore and South Korea and the Millennials population in Singapore and South Korea. Chapter 5 on Research Design and Methodology, discusses the formulation of the research design and adoption of the appropriate research method. This includes the formulation of the survey questionnaire, the method of analyzing the data using descriptive and inferential statistics as well as providing qualitative inputs of the interview findings and background information of the interviewees. Chapter 6 on Research Findings and Analysis presents findings of the descriptive and inferential statistics. In the descriptive statistics, the mean scores computed from the Singapore and South Korea survey results are ranked and compared individually to evaluate which attribute has the highest or lowest impact to the individual. The mean differences between the Singapore and South Korea survey results are also ranked and compared. This identifies the clear distinction that separates the characteristics of the Singaporean and South Korean respondents. The inferential statistics computed from the Singapore and South Korea surveys are used to explain the conceptual framework. Structural equation modelling is chosen to perform multivariate analysis on the conceptual framework as this method simultaneously examines the relationships between variables. A series of computation will be conducted to assess the validity and reliability of the measurement model as well as testing the hypotheses of the conceptual framework by assessing the structural model. Interviews are also conducted to gather the interviewees' views and opinions of the findings from the statistical analysis. Chapter 7 concludes this book by discussing the theoretical and practical implications, strengths and limitations of this study as well as recommendations for further study.

References

Bannon, S., Ford, K., & Meltzer, L. (2011). Understanding millennials in the workplace. *CPA Journal, 81*(11), 61–65.

Carlson, D. S., Grzywacz, J. G., & Zivnuska, S. (2009). Is work–family balance more than conflict and enrichment? *Human Relations, 62*(10), 1459–1486. https://doi.org/10.1177/0018726709336500.

Grzywacz, J. G., & Carlson, D. S. (2007). Conceptualizing work family balance: Implications for practice and research. *Advances in Developing Human Resources, 9*(4), 455–471. https://doi.org/10.1177/1523422307305487.

Lingard, H., & Francis, V. (2009). *Managing WLB in construction*. Hoboken: Taylor and Francis.

Ministry of Manpower. (2011). *Law protects employees from working excessive hour*. Retrieved June 1, 2017 from http://www.mom.gov.sg/newsroom/press-replies/2011/law-protects-employees-from-working-excessive-hour.

Ministry of Manpower. (2016). *Labour market 2015*. Singapore: Ministry of Manpower.

Ofori, G. (2000). Challenges of construction industries in developing countries: Lessons from various countries. Retrieved from https://www.irbnet.de/daten/iconda/CIB8937.pdf

Rantanen, J., Kinnunen, U., Mauno, S., & Tillemann, K. (2010). Introducing theoretical approaches to WLB and testing a new typology among professionals. *Creating Balance?* 27–46. https://doi.org/10.1007/978-3-642-16199-5_2.

Chapter 2
Work–Life Balance and Work–Life Interface

Abstract This chapter provides insights on the existing Work–life Balance (WLB) literature as well as discusses the Work–Life Interfaces associated with WLB. This chapter fundamentally presents the literature review associated with WLB as well as developing the background knowledge pertinent to WLB research essential for developing the conceptual framework presented in the next chapter.

2.1 Overview of the Chapter

This chapter discusses the theory of Work–Life Balance based on past research studies as well as reviews all the different Work–Life Interface theories associated with Work–Life Balance. There are six Work–Life Interface models: Boundary, Spillover, Compensation, Segmentation-Integration, Work–Life Enrichment and Work–Life Conflict.

2.2 Work–Life Balance

Work–Life balance (WLB) is increasingly being recognized as one of the major issues facing today's workforce (Hall & Richter, 1988). Societal changes both globally and locally have created an impact on the individual's lifestyle (Haddon, Hede, & Whiteoak, 2009). Some of the issues that affect the concept of WLB includes: diversified workforce due to demographic shifts and communication technology causing boundaries between work and personal life to blur (Poelmans, Kalliath, & Brough, 2008). These current issues which affect WLB are totally different from the past. Hence, many research studies on WLB completed in the 1980s and 1990s may not be applicable in today's context.

Past research studies on WLB and Work-Family Balance (WFB) have described the same phenomenon; WLB is more comprehensive than WFB as it defines more

L. Sui Pheng and B. K. Q. Chua, *Work–Life Balance in Construction*,
Management in the Built Environment,
https://doi.org/10.1007/978-981-13-1918-1_2

personal roles (friend, mentor, neighbor, etc.) than just the family role (Rantanen, Kinnunen, Mauno, & Tillemann, 2010).

It appears that WLB is an underdeveloped concept today despite its widespread mention in existing work–life and work-family literature (Carlson, Grzywacz, & Zivnuska, 2009; Grzywacz & Carlson, 2007; Poulose & Sudarsan, 2014). To-date there is still a lack of consensus and consistency as to how WLB is to be defined, measured and researched (Rantanen et al., 2010). Similarly, Poelmans, Kalliath, & Brough (2008) also mentioned that existing studies on WLB are fragmented and often inconsistent, making comparison problematic. There are several definitions of WLB based on past completed studies:

(a) Marks and Macdermid (1996: 421) interpreted WFB as role balance where it is defined as "the tendency to become fully engaged in the performance of every role in one's total role system, to approach every typical role and role partner with an attitude of attentiveness and care."
(b) Clark (2000: 751) has defined WFB as "satisfaction and good functioning at work and at home with a minimum of role conflict."
(c) Greenhaus, Collins, and Shaw (2003: 513) modified Marks and MacDermid (1996) and defined WFB as "the extent to which an individual is equally engaged in and equally satisfied with his or her work role and family role."
(d) Frone (2003) defined WFB as a state wherein an individual's work and family life experienced little conflict while enjoying substantial facilitation.
(e) Voydanoff (2005: 825) defined WFB as "global assessment that work and family resources are sufficient to meet work and family demands such that participation is effective in both domains."
(f) Grzywacz and Carlson (2007: 455) defined WFB as "accomplishment of role-related expectations that are negotiated and shared between and individual and his or her role-related partners in work and family domains."

The absence of a strong theoretical understanding of WLB has resulted in literature driven by diverse empirical findings and loosely connected by notions associated to WLB (Carlson, Grzywacz, & Zivnuska, 2009). Moreover, the absence of a conceptually based measurement for researchers to document the level of WLB has also impaired the ability to evaluate sustainable strategies in promoting WLB (Carlson et al. 2009). However, WLB is still an important area of concern that has become part of our daily language which has meaning to each of the individuals (Poelmans, Kalliath, & Brough, 2008).

Many past studies have observed that a favorable work environment which promotes WLB will benefit both the employee and the employer (Baltes, Clark, & Chakrabarti, 2009). Grzywacz and Carlson (2007) also mentioned that WLB contributes to individual health and well-being and is vital for a well-functioning society. Indicators of WLB (work domain) have been heavily linked with employee retention, job satisfaction and individual commitment to the organization (Grzywacz & Carlson, 2007). Likewise, a greater WLB (family domain) is associated with better marital, family satisfaction and better family performance (Carlson, Grzywacz, & Zivnuska, 2009).

2.3 Work and Life Interrelationship

Voydanoff (2005) uses the Ecological Systems Theory and the Boundary Theory to relate the interrelationships between work and life domains. The Ecological Systems Theory explains that a person's development is affected by everything in the surrounding environment (Bronfenbrenner, 1989). Work and life domains belong to the microsystem (Voydanoff, 2005). It is the most influential level in the Ecological Systems Theory as these are the ones who are the closest and have direct contact with the individual (Bronfenbrenner, 1989). Therefore, work and life domains are seen to have a direct impact and influence on an individual's WLB. Relationships and interactions between work and life domains are called mesosystem. Mesosystem explains that individual's roles do not function independently but are interconnected (Bronfenbrenner, 1989). This concurred with past studies that opined work and life are interrelated domains and implied that negative or positive experience in one domain may have significant impact in other domains (Voydanoff, 2005; Poulose & Sudarsan, 2014). The level of influence between each domain can be controlled by individual's boundary management. The Boundary Theory explains this phenomenon and will be discussed later. Furthermore, understanding the causal relationship between work and life domains through these two theories will help in the analysis of the WLB influence on an individual. Voydanoff (2002) termed this causal relationship as Work-Family Interface (WFI) and proposed several mechanisms from past research studies to link work and family characteristics.

2.4 Work–Life Interface

As mentioned earlier, Work–Life Interface (WLI) is seen as an important tool in analyzing WLB. WLI explains the interaction between work and life relationship through integrating different theoretical concepts related to Work–Life Balance. Work and family domain are not seen as a fully separated sphere but as an interdependent domain with permeable boundaries (Desrochers & Sargent, 2004). WLI has long been guided by role theory where individuals hold more than one role with self and socially imposed expectations and appropriate behaviors to these positions (Jang & Zippay, 2011). Within these roles, individuals may encounter conflicts when one role does not meet the expectation of the other (Greenhaus & Beutell, 1985; Jang & Zippay, 2011). This is termed as Work–Life Conflict (WLC). Greenhaus and Powell (2006: 73) defined the positive relationship between work and family as "the extent to which experiences in one role improve the quality of life in another". This relationship is known as Work–Life Enrichment (WLE).

It is important to review the core Work-Family Interface theories as these provide a mechanism that links the relationship between work and family domains (Edwards & Rothbard, 2000). Zedeck and Mosier (1990) have consolidated theories from past research studies and categorized their findings into five models: Spillover, Compensation, Segmentation-Integration, Work–Life Enrichment and Work–Life Conflict. Furthermore, research completed by Basile (2014) included the Border Theory as a potential explanatory mechanism which could alter the outcomes of WLB. All six models must be understood as a whole in order for us to form a basic understanding of WLI which eventually leads to the understanding of WLB.

2.4.1 *Spillover and Compensation*

Spillover refers to the "effects of work and family on one another that generate similarities between the two domains (Edwards & Rothbard, 2000: 180). It is an important linkage between work and family domains in this contemporary research (Grzywacz, Almeida, & Mcdonald, 2002). Similar to WLC and WLE, spillover can occur in both directions (Bell, Rajendran, & Theiler, 2012; Basile, 2014) and it can be in the form of negative or positive spillover (Sumer & Knight, 2001; Grzywacz, Almeida, & Mcdonald, 2002). However, the direction of the spillover was found to be dependent on the more salient role which individual chooses and the magnitude of negative sanctions due to non-compliance (Greenhaus & Beutell, 1985; Basile, 2014). Compensation refers to the efforts to offset dissatisfaction or negative experiences in one domain by seeking satisfaction or positive experiences from another domain (Edwards & Rothbard, 2000). Compensation is an important mechanism in supporting the role of an individual by supplementing positive experience into the affected role domain (Basile, 2014).

Based on the literature (Poelmans, Kalliath, & Brough, 2008; Grzywacz, Almeida, & Mcdonald, 2002; Bell, Rajendran, & Theiler, 2012), it can be seen that the spillover and compensation concept is highly associated with WLC, WLE and WLB. Hence, it appears that spillover and compensation directly affect WLC and WLE without using the boundary/border theory as an intermediary.

2.4.2 *Boundary Theory*

When studying the work–life interface, the boundaries around each domain is another critical factor to understand how multiple roles can lead to conflicts or enrichment (Basile, 2014). This interface is known as the boundary theory or border theory (Basile, 2014; Desrochers & Sargent, 2004). The border theory was proposed by Clark (2000: 750) who defined it as "how individuals manage and negotiate the work and family spheres and the borders between them to attain balance", while the boundary theory is defined as "the way in which individuals create and maintain boundaries as a means of simplifying and ordering the environment". Even though the border theory and the boundary theory are defined differently, the two theories appear to share the same set of propositions

(Desrochers & Sargent, 2004). The boundary works in two ways; i.e. the boundary for an individual transiting from work to life domain or vice versa. It is obvious that work to life domain is easier to control than life to work domain. Organizational policies and culture are seen as critical factors in controlling the transition and boundary between these two domains. Having a policy that restricts employees from communicating with their family at work can block the transition, making the boundary clear. On the other hand, individuals are able to control their own personal life making the boundary blurry as they can transit life to work domain whenever they want. Hence, psychological and behavioral traits play a part in assessing such transitions.

Both theories study the work-family linkages by describing the conditions which work-family integration can improve or diminish individual wellbeing (Desrochers & Sargent, 2004). Furthermore, it addresses how individuals construct, maintain, negotiate and cross boundaries or borders via the line of demarcation (Clark, 2000; Desrochers & Sargent, 2004).

Based on the boundary/border theory, researchers have demonstrated that individuals have varying degrees to maintaining the boundaries between work and non-work domain (Basile, 2014). This variation depends on the ability of the individual's boundary to prohibit or allow the flow from one direction to another (Ashforth, Kreiner, & Fugate, 2000; Clark, 2000). Some individuals may maintain high separation between domains with little spillover while others may integrate both domains during work or non-work period (Basile, 2014). As such, the theory views the relationship between work and home domain as a continuum ranging from segmentation to integration (Baltes, Clark, & Chakrabarti, 2009; Kossek & Lautsch, 2012).

2.4.3 Segmentation-Integration

Boundaries can be managed by either segmentation or integration (Basile, 2014). Segmentation is the separation of the work and family domains such that one domain does not affect the other (Edwards & Rothbard, 2000). This means the boundary is clear and easier to maintain when roles are separated within the line of demarcation (Ashforth, Kreiner, & Fugate, 2000; Desrochers & Sargent, 2004). Integration happens when two or more roles are flexible and permeable (Clark, 2000; Desrochers & Sargent, 2004). This will help to facilitate the transition between the two domains making the boundary blurry (Baltes, Clark, & Chakrabarti, 2009; Ashforth, Kreiner, & Fugate, 2000). It is uncommon to have individuals display complete segmentation or complete integration (Baltes, Clark, & Chakrabarti, 2009). Most individuals will experience less extreme sides of integration or segmentation in order to reap the benefits from each end of the continuum (Baltes, Clark, & Chakrabarti, 2009). The management of each domain and the enactment of segmentation or integration are dependent on the personal and environmental factors (Basile, 2014). Hence, it is said that the organizational factor

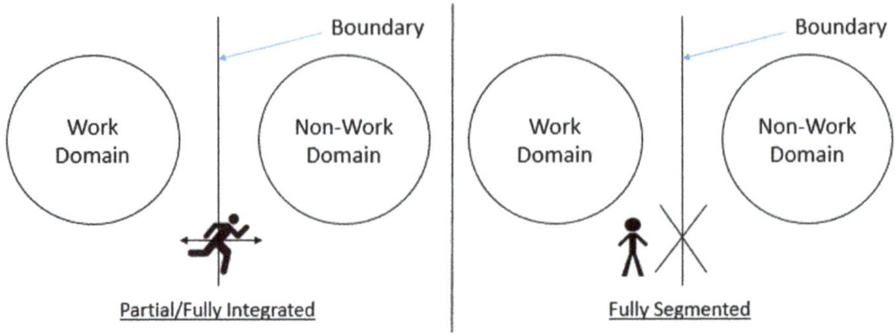

Fig. 2.1 Model showing partial/fully integrated domain and fully segmented domain

plays an important role in determining which continuum the employee/individual will choose (Baltes, Clark, & Chakrabarti, 2009; Mellner, Aronsson, & Kecklund, 2015). Employees who are given flexible work timings or work at home arrangements will have high level of integration. Similarly, organizational policies such as no personal matters are allow at the work place will cause employees to have high level of segmentation. As such, an organization could adapt to the needs of each employee based on their demands in other areas of life or at different life stages so as to allow employees to manage their boundary style which best suits them (Mellner, Aronsson, & Kecklund, 2015). An example relating to the integration and segmentation of an individual developed for this study is shown in Fig. 2.1.

2.4.3.1 Permeability-Flexibility

The Boundary/Border theory posits that each individual's choice of segmentation or integration depends on the extent of flexibility and permeability of the boundaries (Mellner, Aronsson, & Kecklund, 2015). Moreover, the boundary strength of each individual is determined by the levels of permeability and flexibility within which the levels will lead to different outcomes (Basile, 2014; Clark 2000; Ashforth, Kreiner, & Fugate, 2000). Permeability is defined as the degree to which elements from other domains may enter (Hall & Richter, 1988). This implies that permeability is a physical or psychological factor which enables one domain to enter into the current domain which an individual is enacting (Basile, 2014). Flexibility is defined as "the extent to which the physical time and location markers, such as working hours and workplace, may be changed" (Hall & Richter, 1988: 215). An example of boundary flexibility happens when an individual is allow to adjust his/her working hours to suit the non-work needs (Basile, 2014). An inflexible approach occurs when work scheduled is fixed and no time allocation is allow for non-work related matters to exist alongside the scheduled work (Baltes, Clark, & Chakrabarti, 2009). Based on these explanations, high boundary permeability and flexibility would produce role integration while inflexible and impermeable

Table 2.1 Relationship between segmentation, integration, permeability and flexibility

	Segmentation (S)	Integration (I)	Permeability (P)	Flexibility (F)
Segmentation (S)	**X**	High S = Low I Low S = High I Full S = No I	High S = Low P Low S = High P Full S = No P	High S = Low F Low S = High F Full S = No F
Integration (I)	High I = Low S Low I = High S Full I = No S	**X**	High I = High P Low I = Low P	High I = High F Low I = Low F
Permeability (P)	High P = Low S Low P = High S	High P = High I Low P = Low I	**X**	High P = High F Low P = Low F
Flexibility (F)	High F = Low S Low F = High S	High F = High I Low F = Low I	High F = High P Low F = Low P	**X**

boundary would produce role segmentation (Ashforth, Kreiner, & Fugate, 2000). Table 2.1 shows the relationship between Segmentation, Integration, Permeability and Flexibility in the context of this study.

2.4.4 Work–Life Conflict

Work–Life Conflict (WLC) is a source of stress individuals may experience throughout their lives (Carlson, Kacmar, & Williams, 2000). This concept, as defined by Greenhaus, & Beutell (1985: 77), is "a form of inter-role conflict in which role pressures from work and family domains are mutually incompatible". The role pressures in this case are bidirectional in nature and produce negative effects from one domain to another (Carlson, Kacmar, & Williams, 2000; Michel, Kotrba, Mitchelson, Clark, & Baltes, 2010). These bidirectional dimensions are termed as work to life conflict and life to work conflict. It is important to understand the concept of WLC in both directions because the sources of conflict and outcomes produced by work to life conflict and life to work conflict are different (Carlson, Kacmar, & Williams, 2000). Furthermore, Greenhaus, & Beutell (1985) identified three forms of WLC as time-based conflict, strain-based conflict and behavior-based conflict. Time-based conflict occurs when the time devoted to one domain makes it difficult for individuals to engage in another domain. Strain-based conflict occurs when increased stress or tension experienced in one domain hinder the performance in other role. Behavior-based conflict occurs when enacting role behavior may be incompatible with the behavioral expectation of another role. Even though bidirectional relationships and the three forms of WFC were widely agreed upon by researchers, many of them have not translated both concepts into measurable scales (Low, Liu, & Leow, 2010). As a result, measurements were inadequate in explaining WFC. Therefore, Carlson, Kacmar, and Williams (2000) developed a multi-dimensional scale that captured the bidirectional relationships and three forms of WLC as summarized in Table 2.2.

Table 2.2 Multidimensional concept of work-family conflict

Forms of work-family conflict	Directions of work-family conflict: Work Interference with Family	Family Interference with Work
Time	Time-based Work Interference with Family	Time-based Family Interference with Work
Strain	Strain-based Work Interference with Family	Strain-based Family Interference with Work
Behavior	Behavior-based Work Interference with Family	Behavior-based Family Interference with Work

Source Carlson, Kacmar, and Williams (2000)

2.4.5 Work–Life Enrichment

Work–Life Enrichment (WLE) represents the positive side of the WLI by providing individuals with resources that may help them perform better in other life domains (Carlson, Kacmar, Wayne, & Grzywacz, 2006). WLE encapsulates bidirectional dimensions where it is defined as the resources gained in one role helping to improve the quality of life in another role (Greenhouse, & Powell, 2006; Carlson et al., 2006). Similar to the concept of WLC, work to life enrichment and life to work enrichment should be studied separately because both dimensions have different indicators and outcomes. Greenhaus and Powell (2006) mentioned that WLE can occur in two pathways; instrumental and affective path. Instrumental path occurs when resources gained in one role directly increase the performance of other roles and affective path occurs when resources gained in one role indirectly increase the performance on other roles. However, Carlson et al. (2006) found that just by understanding the two paths are not enough to substantiate the occurrence of WLE. This is because there are wide arrays of potential resources that can be transferred along the two paths and are empirically distinct from each other (Carlson et al., 2006). Hence, consistent with the notion of instrumental and affective paths, Carlson et al. (2006) developed and validated four dimensions or "types" of resource gains to measure WLE; namely developmental, efficiency, affective and capital gains. Each direction of WLE is represented by three (out of four) dimensions of resource gains which are shown in Table 2.3. Examples of the four types of resource gains are shown in Table 2.4.

WLE represents the positive side of WLI while WLC represents the negative side of WLI. However, both dimensions have distinct determinants and outcomes

Table 2.3 Three dimensions representing each direction of WLE

Work to family enrichment	Family to work enrichment
Development	Development
Affect	Affect
Capital	Efficiency

Source Carlson, Kacmar, Wayne, and Grzywacz (2006)

Table 2.4 Example of the types of gains associated to WLE

Types of gain	Example
Developmental	Acquisition of knowledge, skills or values
Affective	Changes in behavior and/or attitudes
Capital	Assets acquisition
Efficiency	Attainment of an increased focus level

Source Mcmillan, Morris, and Atchley (2010)

(Haddon, Hede, & Whiteoak, 2009) which coincide with the literature on WLC and WLE mentioned earlier. Hence, it is important to note that WLE and WLC are regarded as an orthogonal construct which means it is possible to experience both positive and negative dimensions of WLI at the same time (Frone, 2003; Poelmans, Kalliath, & Brough, 2008; Rantanen et al., 2010). A good example to explain this phenomenon is when an individual can experience a job with high role conflict in the form of high stress level and long work hours but at the same time, providing a high degree of role enrichment in the form of financial security and personal growth (Grzywacz & Marks, 2000).

Empirical evidence has suggested that linking mechanisms associated to WLI must operate simultaneously in order to provide a useful conceptual basis for comprehending the dynamics of WLB (Frone, 2003). As such, the conceptual framework formulated for this study (Chap. 3) has taken into account the various WLI that is necessary to operationalize the concept of WLB.

2.5 Summary

This chapter reviews the past research problems associated with the WLB concept. In order to understand the concept of WLB comprehensively, one should not look at just the WLB concept but rather concepts that are associated with it. WLI is an important tool to explain the interaction between work and life relationship through integrating different theoretical concepts related to WLB. All six models associated with WLI should be considered synergistically as a whole.

References

Ashforth, B. E., Kreiner, G. E., & Fugate, M. (2000). All in a day's work: Boundaries and micro role transitions. *The Academy of Management Review, 25*(3), 472. https://doi.org/10.2307/259305.

Baltes, B. B., Clark, M. A., & Chakrabarti, M. (2009). WLB: The roles of work-family conflict and work-family facilitation. *Oxford Handbooks Online*. https://doi.org/10.1093/oxfordhb/9780195335446.013.0016.

Basile, K. (2014). Framing the work-life relationship—Understanding the role of boundaries, context and fit (Doctoral dissertation). Retrieved from http://etheses.lse.ac.uk/892/1/Basile_Framing%20the%20work-life%20relationship.pdf.

Bell, A. S., Rajendran, D., & Theiler, S. (2012). Job stress, wellbeing, WLB and work-life conflict among Australian academics. *EJAP E-Journal of Applied Psychology, 8,* 1. https://doi.org/10.7790/ejap.v8i1.320.

Bronfenbrenner, U. (1989). Ecological systems theory. *Annals of Child Development, 6,* 187–249.

Carlson, D. S., Grzywacz, J. G., & Zivnuska, S. (2009). Is work–family balance more than conflict and enrichment? *Human Relations, 62*(10), 1459–1486. https://doi.org/10.1177/0018726709336500.

Carlson, D. S., Kacmar, K. M., Wayne, J. H., & Grzywacz, J. G. (2006). Measuring the positive side of the work–family interface: Development and validation of a work–family enrichment scale. *Journal of Vocational Behavior, 68*(1), 131–164. https://doi.org/10.1016/j.jvb.2005.02.002.

Carlson, D. S., Kacmar, K., & Williams, L. J. (2000). Construction and initial validation of a multidimensional measure of work-family conflict. *Journal of Vocational Behavior, 56*(2), 249–276. https://doi.org/10.1006/jvbe.1999.1713.

Clark, S. C. (2000). Work/family border theory: A new theory of work/family balance. *Human Relations, 53,* 747–770.

Desrochers, S., & Sargent, L. D. (2004). Boundary/border theory and work-family integration 1. *Organization Management Journal, 1*(1), 40–48. https://doi.org/10.1057/omj.2004.11.

Edwards, J. R., & Rothbard, N. P. (2000). Mechanisms linking work and family: Clarifying the relationship between work and family constructs. *The Academy of Management Review, 25*(1), 178. https://doi.org/10.2307/259269.

Frone, M. R. (2003). Work–family balance. In J. C. Quck & L. E. Tetrick (Eds.), *Handbook of occupational health psychology* (pp. 13–162). Washington, DC: American Psychological Association.

Greenhaus, J. H., & Beutell, N. J. (1985). Sources of conflict between work and family roles. *Academy of Management Review, 10*(1), 76–88. https://doi.org/10.5465/amr.1985.4277352.

Greenhaus, J. H., Collins, K. M., & Shaw, J. D. (2003). The relation between work–family balance and quality of life. *Journal of Vocational Behavior, 63*(3), 510–531. https://doi.org/10.1016/s0001-8791(02)00042-8.

Greenhaus, J. H., & Powell, G. N. (2006). When work and family are allies: A theory of work-family enrichment. *Academy of Management Review, 31*(1), 72–92. https://doi.org/10.5465/amr.2006.19379625.

Grzywacz, J. G., Almeida, D. M., & Mcdonald, D. A. (2002). Work-family spillover and daily reports of work and family stress in the adult labor force. *Family Relations, 51*(1), 28–36. https://doi.org/10.1111/j.1741-3729.2002.00028.x.

Grzywacz, J. G., & Carlson, D. S. (2007). Conceptualizing work family balance: Implications for practice and research. *Advances in Developing Human Resources, 9*(4), 455–471. https://doi.org/10.1177/1523422307305487.

Grzywacz, J. G., & Marks, N. F. (2000). Reconceptualizing the work–family interface: An ecological perspective on the correlates of positive and negative spillover between work and family. *Journal of Occupational Health Psychology, 5*(1), 111–126. https://doi.org/10.1037/1076-8998.5.1.111.

Haddon, B., Hede, A., & Whiteoak, J. (2009). *WLB: Towards an integrated conceptual framework.* Retrieved from http://www.anzam.org/wp-content/uploads/pdf-manager/1294_HADDON_BARBARA-362.PDF.

Hall, D. T., & Richter, J. (1988). Balancing work life and home life: What can organizations do to help? *Academy of Management Executive, 2*(3), 213–223. https://doi.org/10.5465/ame.1988.4277258.

Jang, S. J., & Zippay, A. (2011). The juggling act: Managing work-life conflict and WLB. *Families in Society: The Journal of Contemporary Social Services, 92*(1), 84–90. https://doi.org/10.1606/1044-3894.4061.

Kossek, E. E., & Lautsch, B. A. (2012). Work-family boundary management styles in organizations: A cross-level model. *Organizational Psychology Review, 2*(2), 152–171. https://doi.org/10.1177/2041386611436264.

Low, S. P., Liu, J. Y., & Leow, W. Y. (2010). Work-family life of consultant quantity surveyors in Singapore. *Journal of Quantity Surveying and Construction Business, 1*(1), 1–23.

Marks, S. R., & Macdermid, S. M. (1996). Multiple roles and the self: A theory of role balance. *Journal of Marriage and the Family, 58*(2), 417. https://doi.org/10.2307/353506.

Mcmillan, H. S., Morris, M. L., & Atchley, E. K. (2010). Constructs of the work/ life interface: A synthesis of the literature and introduction of the concept of work/life harmony. *Human Resource Development Review, 10*(1), 6–25. https://doi.org/10.1177/1534484310384958.

Mellner, C., Aronsson, G., & Kecklund, G. (2015). Boundary management preferences, boundary control, and WLB among full-time employed professionals in knowledge-intensive, flexible work. *Nordic Journal of Working Life Studies NJWLS, 4*(4), 7. https://doi.org/10.19154/njwls.v4i4.4705.

Michel, J., Kotrba, L., Mitchelson, J., Clark, M., & Baltes, B. (2010). Antecedents of work-family conflict: A meta-analytic review. *Journal of Organizational Behavior, 32*(5), 689–725. https://doi.org/10.1002/job.695.

Poelmans, S. A., Kalliath, T., & Brough, P. (2008). Achieving WLB: Current theoretical and practice issues. *Journal of Management and Organization J Man Org, 14*(3), 227–238. https://doi.org/10.5172/jmo.837.14.3.227.

Poulose, S., & Sudarsan, N. (2014) Work life balance: A conceptual review. *International Journal of Advances in Management and Economics.*

Rantanen, J., Kinnunen, U., Mauno, S., & Tillemann, K. (2010). Introducing theoretical approaches to WLB and testing a new typology among professionals. *Creating Balance?* 27–46. https://doi.org/10.1007/978-3-642-16199-5_2.

Sumer, H. C., & Knight, P. A. (2001). How do people with different attachment styles balance work and family? A personality perspective on work-family linkage. *Journal of Applied Psychology, 86*(4), 653–663. https://doi.org/10.1037/0021-9010.86.4.653.

Voydanoff, P. (2002). Linkages between the work-family interface and work, family, and individual outcomes: An integrative model. *Journal of Family Issues, 23*(1), 138–164. https://doi.org/10.1177/0192513x02023001007.

Voydanoff, P. (2005). Toward a conceptualization of perceived work-family fit and balance: A demands and resources approach. *Journal of Marriage and Family, 67*(4), 822–836. https://doi.org/10.1111/j.1741-3737.2005.00178.x.

Zedeck, S., & Mosier, K. L. (1990). Work in the family and employing organization. *American Psychologist, 45*(2), 240–251. https://doi.org/10.1037/0003-066x.45.2.240.

Chapter 3
Constructing the Conceptual Framework

Abstract This chapter presents a conceptual framework based on the WLB concepts and theories drawn from past research studies. This chapter also explains each of the indicators used in the conceptual framework in details. This chapter presents the research hypotheses to show the relationships between the succeeding and preceding attributes.

3.1 Background in Constructing the Conceptual Framework

The conceptual framework developed in this research is a coalescence of all significant theoretical and conceptual implications of WLB. Many notable research studies do not include empirical findings relating to their respective conceptual frameworks that may in turn affect its validity. To ensure that the conceptual framework developed in this current study have a strong foundation in operationalizing WLB, the proposed model would be backed by conclusive empirical findings from past research studies as well as new findings from this present study. Figure 3.1 shows the conceptual framework designed for this current study. Each of the indicators presented would be explained in this chapter. The research hypotheses presented serve to provide a better understanding of the path relationships between the succeeding and preceding attributes.

There are two approaches to assessing WLB; by using the overall appraisal approach or the components approach (Grzywacz & Carlson, 2007). The overall appraisal approach refers to the general assessment of their WLB by individuals (Rantanen, Kinnunen, Mauno, & Tillemann, 2010). In this approach, WLB is assessed using only general questions like "All in all how successful do you feel in balancing your work and personal life?" (Rantanen et al., 2010: 30). As WLB can vary substantially between individuals, it is hard to assess what causes individual's Work–Life Balance or Imbalance if the overall appraisal approach is used (Grzywacz & Carlson, 2007). The components approach emphasizes WLB as a formative latent construct where WLB is seen as a model which is formed by its

L. Sui Pheng and B. K. Q. Chua, *Work–Life Balance in Construction*,
Management in the Built Environment,
https://doi.org/10.1007/978-981-13-1918-1_3

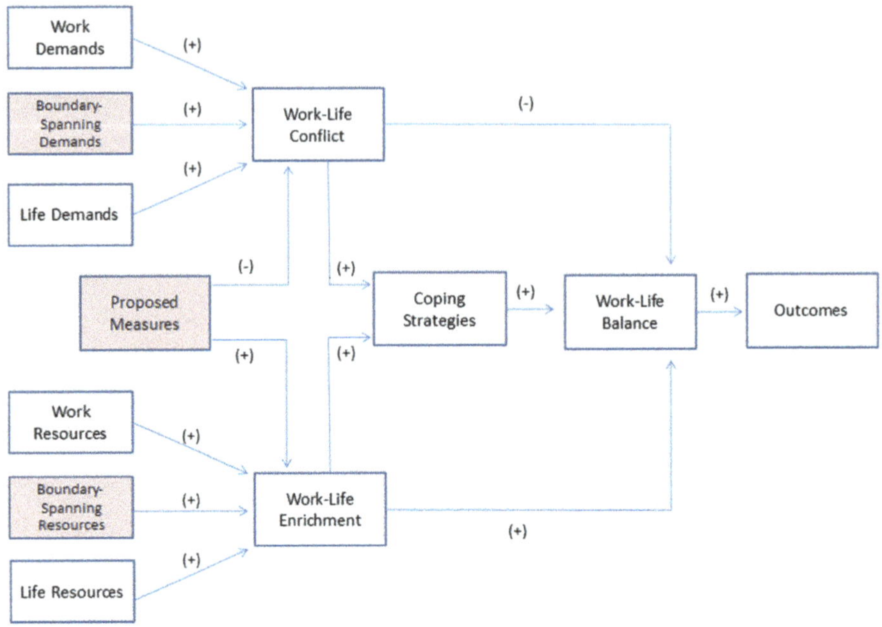

Fig. 3.1 The conceptual framework

antecedent indicators (Edwards & Bagozzi, 2000). This means there are sets of variables that precede and contribute to the outcome and understanding of WLB (Grzywacz & Carlson, 2007; Rantanen et al., 2010). The main advantage of using the components approach is that by using a conceptualized based measurement, it allows researchers to analyze causal relationship of WLB more comprehensively (Rantanen et al., 2010). Furthermore, Grzywacz and Carlson (2007) found that the components approach produced a higher explanation results than overall appraisals. Both the overall approach and the components approach would be used in this current research to cover the overall experience of an individual as well as to use formative latent constructs to conceptualize and measure WLB.

Another related issue lies in the practicality of empirical studies on WLB which may cause results to provide a one-sided view (Grzywacz & Carlson, 2007). One such concern happens when scholars and researchers tend to overemphasize balance as a psychological factor without capturing the social and contextual aspects of WLB (Rantanen, Kinnunen, Mauno, & Tillemann, 2010). This means that only individuals can manage their own WLB and there is little that can be done to help them achieve WLB (Grzywacz & Carlson, 2007). Therefore, Grzywacz and Carlson (2007) proposed that the social perspective must be included to take WLB meaning beyond the individual's perspective. In the conceptual framework proposed in this current study, both psychological and social factors would be analyzed to encompass a more comprehensive view of WLB.

The proposed conceptual framework has documented that demands and resources are bidirectional and are enacted in both work and life domains. The result from demands and resources would directly affect WLB through the appraisal of WLC and WLE. This concurred with the postulations of Frone (2003) as well as Haddon and Hede (2010) where they revealed that low level of WLC and high level of WLE leads to WLB. The level of WLC is positively associated with an individual's ability to meet the demands whereas the level of WLE is positively associated with the amount of resources that an individual possesses to meet the demands and personal performance (Haddon, Hede, & Whiteoak, 2009; Haddon & Hede, 2010).

3.2 List of Indicators in the Conceptual Framework and Hypotheses

3.2.1 Demands and Resources

Based on the proposed conceptual framework, demands and resources are the root causes that have to be examined to explain the outcome of WLC, WLE and WLB. Demands are positively related to WLC and these are the antecedents of WLC (Haddon & Hede, 2010). Demands are defined as "structural or psychological claims to which individuals must respond or adapt by exerting physical or mental effort (Voydanoff, 2004: 398, 2005b: 823). Resources on the other hand are positively related to WLE and these are the antecedents of WLE (Haddon & Hede, 2010). Resources are defined as "structural or psychological assets that are used to facilitate performance, reduce demands or generate additional resources" (Voydanoff, 2004: 398–399, 2005b: 823). The proposed conceptual framework follows the observations made by Voydanoff (2004), Haddon and Hede (2010) as well as Haddon, Hede, and Whiteoak (2009) where demands and resources are differentially salient to WLC and WLE respectively. Since WLC and WLE are seen as independent constructs, this too justifies that demands and resources are not associated with one another. Therefore, increasing individual's work and life resources does not necessarily mean that it is adequate to meet the life and work demands and which in turn reduce WLC and vice versa. Demands and resources in this current research follow Voydanoff's (2005a, b) work where these have been categorized into within-domain demands and resources and boundary-spanning demands and resources.

3.2.1.1 Within-Domain Demands and Resources

(i) **Work Demands and Life Demands**

Work and life demands are characteristics of one domain with processes that limit the ability of an individual to fulfill his/her obligations in another domain

(Voydanoff, 2004, 2005b). Since WLC is a bidirectional concept, work demands are associated with work to life conflict and life demands are associated with life to work conflict (Voydanoff, 2005b). Work and life demands are associated with the antecedents of WLC. Indicators such as work role conflict, work role ambiguity and work involvement are used to measure work demands. Likewise, life role conflict, life role ambiguity and life involvement are used to measure life demands (Carlson, Kacmar, & Williams, 2000; Michel, Kotrba, Mitchelson, Clark, & Baltes, 2010).

(ii) **Work Resources and Life Resources**

Work and life resources from the domain of origin generate positive psychological spillover that enhances the performance of an individual in other domains (Voydanoff, 2005a, b). Work resources are associated with work to life enrichment and life resources are associated with life to work enrichment (Voydanoff, 2005b). Work and life resources are associated with the antecedents of WLE. Indicators such as work salience, developmental experiences and autonomy are used to measure work resources and life salience. Similarly, life harmony is used to measure life resources (Carlson, Kacmar, Wayne, & Grzywacz, 2006). In order to match the number of indicators in the work domain resources, the life rewards indicator proposed by Voydanoff (2005b: 826) was added into the life domain resources.

3.2.1.2 Boundary-Spanning Demands and Resources

(iii) **Boundary Spanning Demands and Resources**

Even though boundary-spanning demands and resources originate from one domain, it has the characteristics that are inherently part of both domains (Voydanoff, 2005b). One example is that of an organization implementing work at home policy to allow employees to take care of their children while working. This policy is seen as a boundary-spanning resource as it enables the individual to enhance his or her WLE. However, as mentioned earlier, the boundary theory is explained by a continuum ranging between segmentation and integration. The degree of flexibility and permeability depends on the level of integration or segmentation an individual is experiencing (Voydanoff, 2005b). For example, if an individual is experiencing high level of segmentation, this means work and life domains are separated and boundaries are impermeable and inflexible (Voydanoff, 2005b). By reiterating this statement, this means boundary-spanning demands and resources influence both domains. However, it is a matter of the individual's boundary strength, either segmentation or integration, that determines the outcome.

The boundary-spanning demand and resource indicators were derived from boundary permeability and flexibility (Voydanoff, 2005a). Boundary permeability is a measurement for boundary spanning demands while boundary flexibility is a measurement for boundary spanning resources (Voydanoff, 2005a). Although the concept of boundary spanning demands and resources are based on Voydanoff's

(2005b) study, not all demands and resources indicators are able to fit into this current research. For instance, Voydanoff's (2005b) indicators focus on the family domain without concern for other non-work domain. In addition, due to the nature of this current research, some of Voydanoff's (2005b) indicators might not be of concern to the respondent, one example of which is "overnight travel". Hence, this study adopted some indicators from Voydanoff (2005b) as well as creating new indicators (Work Domain: Overtime; Life Domain: Participate in workshops for personal development and assigned time for work purposes) as shown in Table 3.1 to fit the aim of this study.

Table 3.1 Indicators for each model of work-life balance

	Work domain	Life domain
(1) Demands	Work role conflict	Life role conflict
	Work role ambiguity	Life role ambiguity
	Work involvement	Life involvement
(2) Resources	Job salience	Life salience
	Developmental experiences	Life harmony
	Autonomy	Life (psychological) rewards
(3) Boundary spanning demands	Work overload	Life activities at work
	Bring work to home	Life interruptions and distraction
(4) Boundary spanning resources	Flexible work schedule	Participate in workshops for personal development
	Time-off for non-Work Purposes	Assigned time for work-purposes
(5) Proposed measures (a) Reduce demands	Take less demanding task	Reduce time with love ones
	No work after 6 pm	Limit usual lifestyles
	Organization plays a part	Reduce role responsibility
(b) Increase resources	Assign more challenging task	Take up new sports/hobby
	More break interval	Plan weekly activity beforehand
	Provide more freedom	Assign more personal time
(6) Work–Life Conflict	Time-based (+)	Time-based
	Strain-based (+)	Strain-based
	Behavior-based (+)	Behavior-based
(7) Work–Life Enrichment	Development	Development
	Affect	Affect
	Capital	Efficiency
(8) Coping strategies	Colleague and employer support	Families and friends support
	Leisure activities	
	Cognitive management	
	Behavioral management	
(9) Outcomes	Job satisfaction	Life satisfaction
	Organizational commitment	Family satisfaction
	Reduce turnover intention	Psychological well-being

The proposed conceptual framework posits two ways that are able to alter the outcomes of WLC and WLE. Firstly, this is by using "Proposed Measures" and secondly by using "Coping Strategies". "Proposed Measures" is a set of determinants recommended by other researchers and in the context of this study are able to help individuals achieved low level of WLC and high level of WLE. This is achieved by reducing the current demands and increasing the current resources available to an individual. "Coping Strategies" is a set of determinants that would alter the actual outcome of WLC and WLE of an individual. For example, if an individual is experiencing high level of WLC, he or she will have ways to manage this impact in order to cope with his or her daily life.

3.2.2 Proposed Measures

It is essential to find out what are the measures that must be undertaken in order to achieve low level of WLC and high level of WLE and if the current demands and resources are not the right avenue to achieving this objective. In this present study, Proposed Measures are used to suggest ways that would enable individual to reduce the current work/family demands and improve work/family resources. This are then tested to determine whether these suggested measures would improve or change the outcome of WLC and WLE. These Proposed Measures coalesce well with Voydanoff's (2005b) work where these measures were termed as "Boundary-Spanning Strategies". Voydanoff (2005b) described these as actions taken by the organization, the family and the individual to help reduce or eliminate misfit between demands and resources (Voydanoff, 2005b). These strategies are categorized into: (a) reduce work/family demands strategies, and (b) increase work/family resources strategies. Each of these strategies have its own set of measurements and scales (Voydanoff, 2005b). Lastly, "Proposed Measures" is negatively associated with WLC and positively associated with WLE.

It is important to note that "Proposed Measures" in the conceptual framework do not alter the outcome of the actual WLC and WLE that an individual is experiencing. These are suggested recommendations that enable organizations or parties concerned to form a foundation that allow them to create policies to help individuals achieve WLB. Although the "Proposed Measures" concept was based on Voydanoff's (2005b) "Boundary Spanning Strategies", most of the work domain indicators suggested were unable to fit into this current research due to the nature of the job of the targeted respondents. Furthermore, the family domain indicators were inadequate for this current study as the research is focusing on life domain. Hence, new indicators (as shown in Table 3.1) were needed for this current research. Most of the life indicators, identified from the work of Lingard and Francis (2009), were modified to match the purpose of this current research. Other indicators identified for this current study includes "no work after 6 pm", "reduce time with loved ones", "limit usual lifestyle" and "plan weekly activity beforehand".

3.2.3 Work–Life Conflict and Work–Life Enrichment

In the proposed conceptual framework, WLE and WLC are viewed as a linking mechanism between resources and demands as well as WLB. These are important indicators for understanding the effects emanating from both directions of demands and resources associated with Work–Life Domains and serving as useful indicators for WLB (Haddon, Hede, & Whiteoak, 2009; Haddon & Hede, 2010; Voydanoff, 2005b). Based on the hypotheses presented below, WLE and WLC are positively associated with resources and demands respectively. It is important to note that WLC and WLE are orthogonal constructs and have distinct determinants and consequences (Haddon, Hede, & Whiteoak, 2009). Hence, the process underlying WLC does not generalize to WLE. Consequently, models that have been used by WLC cannot be substituted for WLE (Frone, 2003). In the proposed model, WLC and WLE operate through two directions, either directly to WLB or using Coping Strategies as a mediator to alter the outcome of WLC or WLE before assessing the corresponding level of WLB. It is postulated that a direct relationship to WLB means individuals have experienced low level of WLC or high level of WLE in their work and life domain without the need to engage in strategies to alter the outcome. On the other hand, operating through Coping Strategies means an individual might experience high level of WLC or low level of WLE. As a result, an individual may engage in strategies that can help alter the outcome of WLC or WLE in order to achieve WLB. This led to the formulation of the following hypotheses:

Hypothesis 1a: Work Demands is positively associated with Work–Life Conflict.
Hypothesis 1b: Boundary Spanning Demands is positively associated with Work–Life Conflict.
Hypothesis 1c: Life Demands is positively associated with Work–Life Conflict.
Hypothesis 2a: Work Resources is positively associated with Work–Life Enrichment.
Hypothesis 2b: Boundary Spanning Resources is positively associated with Work–Life Enrichment.
Hypothesis 2c: Life Resources is positively associated with Work–Life Enrichment.

3.2.4 Coping Strategies

In addition to having a direct relationship with WLB, WLC and WLE may operate through Coping Strategies (CS) with respect to WLB. The proposed model assumes that WLC and WLE have already been experienced by individuals and it is possible to uncover the relevant strategies that individuals adopt to alter the outcome of WLC and WLE in order to successfully balance their work and life. Past research has showed that even if high WLC is the consequence of engaging in work or family roles, it is how individuals cope with the WLC that determine their personal outcomes (Haddon, Hede, & Whiteoak, 2009; Haddon & Hede, 2010). Haddon and

Hede (2009) revealed that despite working long hours and conflicting role demand, employees might still have the perception that they achieved WLB. Frone (2003) mentioned that individuals might seek social support or developed appropriate mentality in order to reduce or better cope with stressors at work and in their life. In addition to personal coping strategies, Voydanoff (2005b) also added that organization and family support also plays a part in helping individual reduce WLC and improve WLE. As such, CS is seen as a mediator for WLC and WLE where individuals could engage in strategies that alter the perception of their achievement in WLB. There are a series of coping strategies proposed by Haddon and Hede (2009) and this study will adopt these strategies as shown in Table 3.1 to assess the effectiveness in altering the level of WLB of an individual. In the context of the models illustrated, this study proposed that WLC and WLE are positively associated to Coping Strategies. This led to the formulation of the following hypotheses:

Hypothesis 3a: Work–Life Conflict is positively associated with Coping Strategies.
Hypothesis 3b: Work–Life Enrichment is positively associated with Coping Strategies.

3.2.5 Work–Life Balance

WLB is seen as a conglomeration of multiple measureable constructs rather than a unidimensional construct (Rantanen et al., 2010). This current study identified past research studies to develop multi-dimensional indicators used to explain WLB of an individual. WLC and WLE are seen as the two important indicators that act as linking mechanisms to explain WLB (Grzywacz & Carlson, 2007). Furthermore, it has been observed that high WLE and/or low WLC will result in achievement of WLB (Frone, 2003). In addition to WLC and WLE as the linking mechanism to WLB, Coping Strategies have been identified as the third way for an individual to achieve WLB. The achievement of WLB is still a debatable issue as there are little or no empirical evidence that was available prior to this current research to justify whether WLE, WLC and Coping Strategies are needed to achieve WLB or either one of these is more significant in explaining WLB. This research will explain the phenomenon using statistical tools to evaluate the model. Finally, WLB is positively associated to its outcomes, which indicates that achieving WLB will yield beneficial consequences to individual's work and life domains (Carlson, Grzywacz, & Zivnuska, 2009; Rantanen et al., 2010). This led to the formulation of the following hypotheses:

Hypothesis 4a: Work–Life Conflict is negatively associated with Work–Life Balance.
Hypothesis 4b: Work–Life Enrichment is positively associated with Work–Life Balance.
Hypothesis 4c: Coping Strategies is positively associated with Work–Life Balance.

3.2.6 Outcomes of Work–Life Balance

It is important to assess the outcomes of WLB to understand the benefits or detrimental effects of work–life balance and work–life imbalance. The findings from Carlson, Grzywacz, & Zivnuska (2009) found that WLB contributes to the explanation of both work and life outcomes. In work outcomes, WLB was posited to improve job satisfaction and organizational commitment and intention to turnover (Carlson, Kacmar, & Williams, 2000; Carlson, Grzywacz, & Zivnuska, 2009). In life outcomes, WLB was posited to improve life satisfaction, family satisfaction and psychological well-being (Carlson et al., 2006). All the indicators associated with work and life outcomes are presented in Table 3.1. This led to the formulation of the following hypothesis:

Hypothesis 5: Work–life balance is positively associated with Outcomes.

3.3 Summary

The main aim of the conceptual framework is to identify the causal mechanism between WLB and its associated theories as well as concepts. The relationships between indicators are based on past studies and the hypotheses were formulated to test their significance to explain WLB.

References

Carlson, D. S., Grzywacz, J. G., & Zivnuska, S. (2009). Is work–family balance more than conflict and enrichment? *Human Relations, 62*(10), 1459–1486. https://doi.org/10.1177/0018726709336500.

Carlson, D. S., Kacmar, K., & Williams, L. J. (2000). Construction and initial validation of a multidimensional measure of work-family conflict. *Journal of Vocational Behavior, 56*(2), 249–276. https://doi.org/10.1006/jvbe.1999.1713.

Carlson, D. S., Kacmar, K. M., Wayne, J. H., & Grzywacz, J. G. (2006). Measuring the positive side of the work–family interface: Development and validation of a work–family enrichment scale. *Journal of Vocational Behavior, 68*(1), 131–164. https://doi.org/10.1016/j.jvb.2005.02.002.

Edwards, J. R., & Bagozzi, R. P. (2000). On the nature and direction of relationships between constructs and measures. *Psychological Methods, 5,* 155–174.

Frone, M. R. (2003). Work–family balance. In J. C. Quck & L. E. Tetrick (Eds.), *Handbook of occupational health psychology* (pp. 13–162). Washington, DC: American Psychological Association.

Voydanoff, P. (2005a). Toward a conceptualization of perceived work-family fit and balance: A demands and resources approach. *Journal of Marriage and Family, 67*(4), 822–836. https://doi.org/10.1111/j.1741-3737.2005.00178.x.

Grzywacz, J. G., & Carlson, D. S. (2007). Conceptualizing work family balance: Implications for practice and research. *Advances in Developing Human Resources, 9*(4), 455–471. https://doi.org/10.1177/1523422307305487.

Haddon, B., & Hede, A. (2009). WLB: In search of effective strategies. Retrieved September 17, 2016 from https://www.researchgate.net/publication/272623357_Work-life_balance_In_search_of_effective_strategies.

Haddon, B., & Hede, A. (2010). WLB—An integrated approach. *University of The Sunshine Coast.*

Haddon, B., Hede, A., & Whiteoak, J. (2009). WLB: Towards an integrated conceptual framework. Retrieved from http://www.anzam.org/wp-content/uploads/pdf-manager/1294_HADDON_BARBARA-362.PDF.

Lingard, H., & Francis, V. (2009). *Managing WLB in construction.* Hoboken: Taylor and Francis.

Michel, J., Kotrba, L., Mitchelson, J., Clark, M., & Baltes, B. (2010). Antecedents of work-family conflict: A meta-analytic review. *Journal Of Organizational Behavior, 32*(5), 689–725. https://doi.org/10.1002/job.695.

Rantanen, J., Kinnunen, U., Mauno, S., & Tillemann, K. (2010). Introducing theoretical approaches to WLB and testing a new typology among professionals. *Creating Balance?,* 27–46. https://doi.org/10.1007/978-3-642-16199-5_2.

Voydanoff, P. (2004). The effects of work demands and resources on work-to-family conflict and facilitation. *Journal of Marriage and Family, 66*(2), 398–412. https://doi.org/10.1111/j.1741-3737.2004.00028.x.

Voydanoff, P. (2005b). Consequences of boundary-spanning demands and resources for work-to-family conflict and perceived stress. *Journal of Occupational Health Psychology, 10*(4), 491–503. https://doi.org/10.1037/1076-8998.10.4.491.

Chapter 4
Singapore and South Korea Context

Abstract This chapter introduces the background of the research scope and comprises of three sections: namely overview of Singapore and South Korea society, the construction industry in Singapore and South Korea and demographics of the Millennials population in Singapore and South Korea.

4.1 Overview of Chapter

This chapter is divided into three sections: Overview of Singapore and South Korea, Construction Industry and Millennials workforce. The overview of Singapore and South Korea explains the background information of both countries by showing the age group distributions of their respective populations as well as presenting the Hofstede's Six Dimension of National Culture for both countries. The Construction Industry (CI) section presents the performance of both countries in terms of GDP growth. In addition, this section also covers the strengths of the workforce, employees' wages and working hours in both countries. Finally, the Millennials section presents the demographic profiles of both countries showing the Millennials population, with comparison of the workforce and education levels.

4.2 Reasons for Comparing Singapore with South Korea

Singapore and South Korea have been labelled as the Asian Tigers from East Asia that exhibited an astonishing rate of economic growth that surpassed many countries in Latin America and Europe (Oh, 1991). East Asian states were typically regarded as a "Confucian society" where the values from Confucianism are believed to be the major contribution towards the radical reforms witnessed in East Asia's economic development (Abe, 2006; Hofstede & Bond, 1988). Today, Confucianism has become an inseparable part of life in the attitudes and behaviors that characterizes the East Asian cultural identity (Oh, 1991). Confucianism is a set

L. Sui Pheng and B. K. Q. Chua, *Work–Life Balance in Construction*,
Management in the Built Environment,
https://doi.org/10.1007/978-981-13-1918-1_4

of pragmatic rules used in daily life without invoking any religious contents (Hofstede, 2001). As noted by Hofstede (2001: 354), there are four key guiding principles in Confucian teachings:

- The stability of society is based on unequal relationships between people. These relationships are based on mutual and complementary obligations.
- The family is the prototype of all social organizations. A person is not primarily an individual; rather, he or she is a member of a family. Harmony is found in the maintenance of everybody's face, in the sense of dignity, self-respect, and prestige.
- Virtuous behavior toward others consists of not treating others as one would not like to be treated oneself.
- Virtue with regard to one's task in life consists of trying to acquire skills and education, working hard, not spending more than necessary, being patient, and persevering. Conspicuous consumption is taboo as is losing one's temper. Moderation is prescribed in all things.

However, Hofstede and Bond (1988) believed that even though East Asia might exhibit similar cultural values, the intensity varies from one country to another. A more comprehensive measurement template termed as Hofstede's Cultural Dimensions was introduced to study the cultural influences on the society with corresponding scores assigned to all six dimensions for each country (Hofstede, 2001). These dimensions are presented later. The numerical score assigned to each dimension allows this current research to compare and contrast the differences between the cultures in Singapore and South Korea.

4.3 Overview of Singapore and South Korea Society

Tables 4.1 and 4.2 show that both Singapore and South Korea are facing an ageing population issue due to the rising proportions of older citizens. The comparison made between 2005 and 2015 on Singapore resident population in Table 4.1 shows that the number of people aged 50 years and above has rose significantly compared to those groups aged 50 years and below. Similarly, the comparison made on the

Table 4.1 Age group distribution for Singapore Resident Population, 2005 and 2015

	0-4	5-9	10-14	15-19	20-24	25-29	30-34
2005	199,529	237,237	258,553	234,921	218,951	247,799	293,414
2015	183,575	204,452	214,388	242,902	264,127	264,127	290,619

	35-39	40-44	45-49	50-54	55-59	60-64	65 and above	Total
2005	299,153	321,472	307,546	254,168	197,803	117,575	279,693	**3,467,814**
2015	301,067	316,755	303,413	315,091	295,063	240,493	459,715	**3,895,787**

Source Department of Statistics Singapore (2016)

Table 4.2 Age group distribution for South Korea Resident Population, 2011 and 2016

	0–4	5–9	10–14	15–19	20–24	25–29	30–34
2011	2,328,435	2,347,414	3,134,531	3,546,594	3,212,315	3,515,101	4,024,454
2016	2,225,116	2,359,473	2,376,960	3,183,878	3,553,038	3,179,502	3,584,698
	35–39	40–44	45–49	50–54	55–59	60–64	65–69
2011	4,230,332	4,617,122	4,229,561	4,313,633	3,214,513	2,319,307	1,897,565
2016	4,002,012	4,251,616	4,559,601	4,169,465	4,229,222	3,055,420	2,228,587
	70–74	75–79	80–84	85–89	90–94	95 and above	Total
2011	1,640,831	1,130,345	618,481	283,425	130,325	35,311	**50,734,284**
2016	1,777,096	1,432,861	886,948	404,079	189,980	50,987	**51,649,552**

Source Korean Statistical Information Service (2016)

South Korean population in 2011 and 2016 shows a significant rise in proportions for those aged 55 years and above. One reason for this phenomenon is that the age group 55–59 years has risen by more than 1 million citizens. On the contrary, most of the age groups below 55 years have declined in numbers. As a result, there is a disproportioned growth between these two categories.

The ageing population trend experienced by Singapore and South Korea has a significant impact on the nation's economy, individual life and organization strategies. One prominent impact on the nation's economy and individual life is the rise of the old-age dependency ratio (Hagemann & Nicoletti, 1989; Pettinger, 2016). Dependency ratio is defined as the ratio of pension-fund supported population to pension-fund supporter (usually working adults aged 20–64 years) (Hagemann & Nicoletti, 1989). A rise in the dependency ratio would affect the funding capability of the government because there would be more old citizens claiming pension and health benefits and lesser working adults paying income taxes (Hagemann & Nicoletti, 1989; Pettinger, 2016). As a result, there would be a need for the government to increase the savings to fund the pension scheme (Hagemann & Nicoletti, 1989; Pettinger, 2016). This would eventually lead to a slower economic growth rate due to the reduction in the nation's capital accumulation (Pettinger, 2016). Hence, one possible solution to increase government funding is to increase the tax contributions from organizations and working adults (Pettinger, 2016).

The practice of increasing tax contributions by the government would have an adverse impact on working adults. It would be a double whammy for the working adults as organizations might cut or stagnant employees' wages in order to match the increasing contributions of tax revenues (Pettinger, 2016). Furthermore, working adults might also be straddled with higher income tax contributions (Hagemann & Nicoletti, 1989). Undoubtedly, this would cause working adults to have greater financial burdens due to the rise in the costs of living while having to bear with reduced or stagnant wages.

Lastly, the rise of an ageing workforce impacts organizations as well. Industries faced high proportions of older workforce while the rate of a younger workforce entering the industry does not match the rate of retirements (Macleod, 2013). This was also experienced in the construction industry. Due to the nature of the construction industry, many organizations faced difficulties attracting a younger workforce to join the industry (Seah, 2013). This would result in a workforce shortage and fall in productivity. Hence, organizations must relook at their hiring policies and strategies to attract and retain a young workforce (Seah, 2013). More details relating to this issue are explained later.

4.4 Hofstede's Cultural Dimensions on Singapore and South Korea

Different cultures in the world have their own unique ways of looking at certain perspectives in societies. Culture is important in allowing us to know how different societies react in a specific situation (Hampden-Turner & Trompenaars, 1998). Hofstede (2001) defined culture as a collective mental programming of the human mind that distinguishes one group of society from another. To understand the cultural influence on human behaviors and values in Singapore and South Korea, Hofstede's cultural dimensions are referred to in this current research.

Hofstede's cultural dimensions provide a framework to compare and contrast between the cultures of different nations (Hofstede, 2001). Hofstede categorized national culture into six dimensions: Power Distance, Individualism/Collectivism, Masculinity/Femininity, Uncertainty Avoidance, Long Term Orientation and Indulgence versus Restraint (Hofstede, Hofstede, & Minkov, 2010). These dimensions help to distinguish one culture from another through their respective scores and evaluating the results on each dimension (Hofstede et al., 2010). Based on the national culture scores as shown in Table 4.3, Singapore and South Korea fared differently in each dimension.

4.4.1 Power Distance

Power Distance (PD) refers to the extent that individuals in the society are not equal and they accept this power difference (Hofstede, 2001). In this dimension, Singapore's score of 74 points signifies that there is a clear power difference and hierarchical order among a social group. South Korea's score of 60 points indicates

Table 4.3 Hofstede's six dimensions of national culture score (over the score of 100)

Countries	Power distance	Individualism versus collectivism	Masculinity versus femininity	Uncertainty avoidance	Long Term orientation	Indulgence versus restraint
Singapore	74	20	48	8	72	46
South Korea	60	18	39	85	100	29

Source Singapore-Geert Hofstede (2016); South Korea-Geert Hofstede (2016)

that South Koreans follow the hierarchical structure present in the society and accept the power inequality among individuals. One obvious reason is due to South Korea's seniority-based culture instilled to the people since young.

4.4.2 Individualism Versus Collectivism

This dimension addresses the degree of interdependency of the members in a society. In an individualistic society, people have an "I" mentality, would only look after themselves and their direct family members (Hofstede, 2001). On the other hand, a collective society has a "we" mentality, would look after one another (social group) in exchange for loyalty (Hofstede, 2001). Collectivism follows the second principle of Confucian teaching where "The family is the prototype of all social organizations. A person is not primarily an individual; rather, he or she is a member of a family (Hofstede, 2001: 354)." In this dimension, Singapore and South Korea scored 18 and 20 points respectively, which mean that both nations belong to the Collectivism societies. This also indicates that Singaporeans and South Koreans exhibit a strong sense of working as a group rather than working individually. They work towards a collective goal where success or failure is shared among the group members and no individual is to be blamed (Hofstede et al., 2010).

4.4.3 Masculinity Versus Femininity

Masculinity refers to "a society in which social gender is clearly distinct: Men are supposed to be assertive, tough and focused on material success; women are supposed to be more modest, tender and concerned with the quality of life (Hofstede, 2001: 297)." Femininity stands for "a society in which social gender roles overlap: Both men and women are supposed to be modest, tender, and concerned with the quality of life (Hofstede, 2001: 297)." Countries with high scores for this dimension are societies driven by competition, achievement and success in life or work. Countries with low scores for this dimension are societies that show care for others and their quality of life. In this dimension, Singapore scored 48 points symbolizing that the Singapore society leans towards the feminine side. This suggests that Singaporeans believe in showing care and empathy to others to improve the unity and quality of life and people. Similarly, South Korea has a low score of 39 points, which suggests the South Koreans possess characteristics of a feminine society.

4.4.4 Uncertainty Avoidance

Uncertainty Avoidance refers to the degree of comfort the members of a society felt when they experienced uncertainty and ambiguity in life (Hofstede, 2001). In this dimension, Singapore scored only 8 points that suggests Singaporeans accept uncertainty and ambiguity in life and they tend to have a more relaxed attitude and are open to changes given to them. Hence, Singaporeans exhibit low stress and anxiety and are more content with life (Hofstede et al., 2010). On the other hand, South Korea has a high score of 85 points that suggests South Koreans value security and conform to formal rules to regulate behavior in the society. In addition, they have a mentality that working hard and following what they were told would lead to a better quality of life. This also means South Koreans exhibit high stress, anxiety and emotions (Hofstede et al., 2010).

4.4.5 Long Term Orientation

Long Term Orientation (LTO) describes how culture values traditions while dealing with the present and future challenges (Hofstede et al., 2010). The LTO dimension reflects traits resembling the teachings of Confucius (Hofstede, 2001). It appears that countries with high LTO scores have high Confucian values and countries with low LTO scores have low Confucian values (Hofstede, 2001). Nations with low LTO scores have normative thinking, have great respect for their traditions and are more resistant to changes or are short sighted. Nations with high LTO scores have a more pragmatic future-oriented perspective and believe that they must always prepare for any changes in the future or are long-sighted. Furthermore, they deem traditions as adaptable to change circumstances (Hofstede et al., 2010). In this dimension, Singapore's score of 72 points suggests Singaporeans have a cultural quality that supports their long-term goals. One such example is Singapore's excellent education system built on preparing the next generation of Singaporeans for the future. South Korea on the other hand has a full score of 100 points in this dimension making them one of the most pragmatic and long-term oriented societies in the world. One example of such a phenomenon are South Korea's companies geared to achieving results in terms of profits and performance for the shareholders. In additions, they also serve the stakeholders and the society for many generations to come.

4.4.6 Indulgence Versus Restraint

Indulgence versus Restraint refers to the degree of control an individual has over his or her life and desires. Society with relatively free gratification of natural human drives related to enjoying life and having fun are associated with Indulgence. Society that suppresses gratification of needs and control by means of strict social norms are associated with Restraint (Hofstede et al., 2010). A preference on Singapore for this dimension could not be determined due to the score (46 points) which is close to a mid-point balance. South Korea's score of 29 points indicates that it belongs to the Restraint culture where South Koreans have the tendency to show cynicism and pessimism towards life. Moreover, they felt restrained by social norms and have a perception that indulging themselves is somewhat inappropriate.

4.5 Overview of the Construction Industry

The construction industry is changing at a rapid pace due to the influence of globalization and technological advancements. The construction industry operates in a highly competitive market with relatively low profit and projects are required to complete within tight deadlines (Lingard & Francis, 2009). In construction contracts, companies are bind by their promises to fulfill all the obligations set out in the agreement. Employees might suffer if there is a project delay because they need to do overtime work. Hence, a combination of long working hours, working with tight schedules and high stress level increases employee's concern over the negative work impacts on personal life (Turner, Lingard, & Francis, 2009). Apart from the potential negative work impacts, construction work have often been perceived as "dirty, difficult and dangerous" (Wells, 2001; Lee, 2014). With these poor images perceived by the potential workforce, employment rate in the construction sector have significantly deteriorated (Wells, 2001).

4.5.1 Singapore's Context

Table 4.4 shows a comparison between the total workforce between 2011 and 2016 in Singapore. Although the workforce increased by around 10,700, the large inclement was geared towards the age group of 50–64 years. This signals that the construction industry is still not attracting the younger people especially from those in the age group of 15–29 years.

Based on the average weekly paid working hours shown in Appendix A, the construction industry has the longest paid working hours as compared to other industries. Even though there was a decrease between 2013 and 2014, having the

Table 4.4 Employed residents from aged 15 to 65 years and above in the Singapore construction industry between 2011 and 2015 (in thousands)

Construction industry	Total	15–19	20–24	25–29	30–34	35–39	40–44
2011	99.7	0.2	1.9	5.3	8.5	12.7	14.3
2015	110.4	0.4	2.7	5.6	9.8	10.2	13.1

Construction industry	45–49	50–54	55–59	60–64	65 and above
2011	19.3	16.2	11.6	5.9	3.7
2015	16.5	20.0	16.9	9.1	6.2

Source Ministry of Manpower (2016a)

longest weekly working hours does not remove the negative image to the potential workforce. Aside from the normal weekly working hours, the construction industry also has the longest weekly paid overtime hours in Singapore as shown in Appendix B.

The construction industry is one of the nine sectors that contribute to Singapore's economy (Ministry of Trade and Industry, 2016). Figure 4.1 shows that the overall GDP in Singapore grew by 2% in 2015. The construction sector has also experienced a growth rate of 2.5, 0.5% more than the overall GDP growth rate. However, the increment was 3.4 and 0.5% less than the growth rate in 2013 and 2014 respectively. This figure could also mean that the growth rate experienced by the construction industry has dropped from 5.9% in 2013 to 3.0% in 2014 to 2.5% in 2015. The reason for this is due to the slowdown in the private sector construction demand (Ministry of Trade and Industry, 2015, 2016).

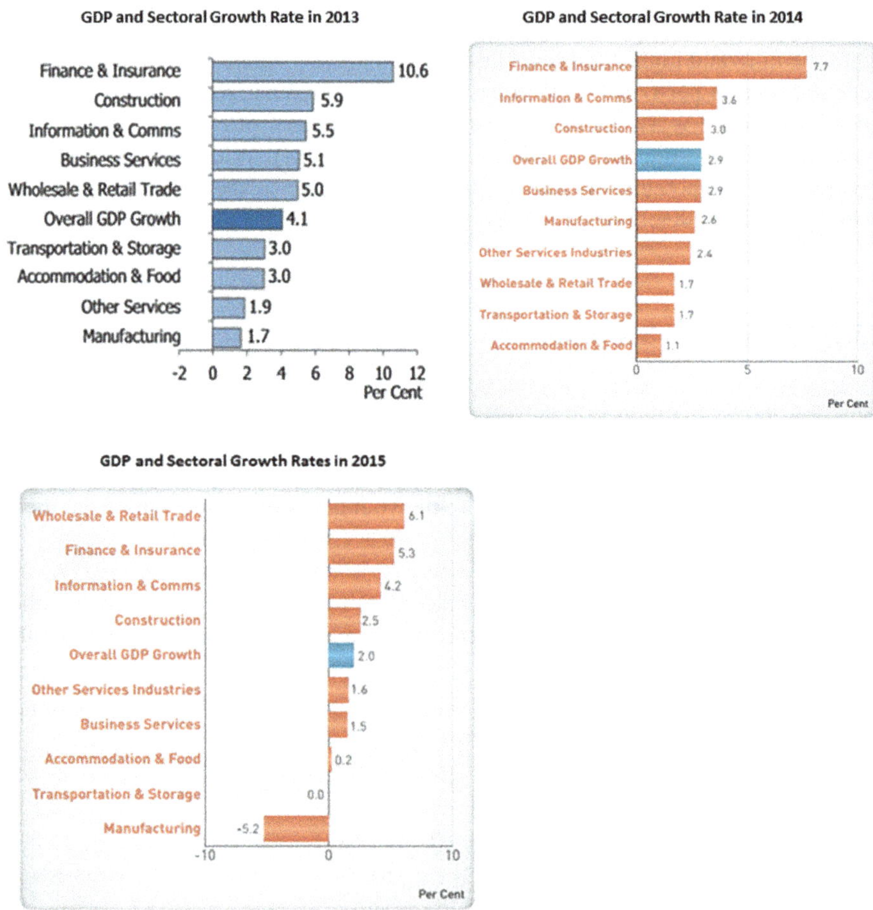

Fig. 4.1 GDP and sectorial growth rates in 2013 (top left), 2014 (top right) and 2015 (bottom). *Source* Ministry of Trade and Industry (2014, 2015, 2016)

Table 4.5 shows that employment in the construction industry constitutes 14% of the total employment in Singapore in 2015. Furthermore, the number of foreign workers in the construction industry is thrice the number of Singaporeans (375,500 vs. 124,500). This number is not surprising given that Singaporeans have always perceived the construction industry as an industry with a high influx of foreign workers (Seah, 2013). However, there was a sharp fall in the number of foreign workers in the construction industry after 2012 partly due to the initiatives taken by the Ministry of Manpower (MOM) to moderate the dominance of foreign workers in the industry. Based on the "Singapore Budget 2012 – Key Budget Initiatives", the MOM has set out new regulations to reduce the dependency on foreign workers in the country. For the construction industry, the Man-Year Entitlement (MYE) quota was reduced which also means that the total number of Work Permit holders a main contractor was entitled to in order to complete the project was also correspondingly reduced.

Table 4.6 presents the median gross monthly income between industries (including the employer's Central Provident Fund (CPF) contributions). Table 4.6 shows that the construction industry has the lowest median gross monthly income per employee in 2014 and 2015. However there was an 8.9% increase from 2014 as

Table 4.5 Employment change by residential status and industry

In Thousands

	Employment Change											Employment Level
	2005	2006	2007	2008	2009	2010	2011	2012	2013	2014	2015	Dec 2015
Overall												
Total	113.3	176.0	234.9	221.6	37.6	115.9	122.6	129.1	136.2	130.1	32.3	3,656.2
Total (excl FDW)	*105.5*	*168.0*	*223.5*	*213.4*	*32.9*	*110.6*	*117.7*	*125.8*	*131.3*	*122.1*	*23.3*	*3,424.7*
Local	63.5	90.9	90.4	64.7	41.8	56.2	37.9	58.7	82.9	96.0	0.7	2,268.9
Foreign	49.8	85.1	144.5	156.9	-4.2	59.7	84.8	70.4	53.3	34.0	31.6	1,387.3
Foreign (excl FDW)	*42.0*	*77.1*	*133.1*	*148.7*	*-8.9*	*54.4*	*79.8*	*67.1*	*48.4*	*26.0*	*22.6*	*1,155.8*
Manufacturing												
Total	29.1	41.6	49.3	19.5	-43.9	-0.8	3.4	11.4	5.3	-4.4	-22.1	513.8
Local	9.7	11.0	7.3	-4.6	-9.5	-0.1	-2.8	2.1	0.9	1.1	-7.3	252.2
Foreign	19.4	30.6	42.0	24.1	-34.3	-0.7	6.1	9.4	4.4	-5.4	-14.8	261.6
Construction												
Total	8.7	20.5	40.4	64.0	24.0	3.4	22.0	39.1	35.2	14.3	8.6	500.0
Local	2.2	5.3	4.4	5.2	4.3	-0.4	2.4	4.2	3.7	4.6	1.8	124.5
Foreign	6.5	15.2	36.0	58.9	19.7	3.8	19.6	34.9	31.6	9.7	6.8	375.5
Services												
Total	73.8	112.7	143.1	136.4	58.6	112.6	96.1	77.0	94.1	119.7	45.5	2,615.2
Total (excl FDW)	*65.9*	*104.7*	*131.7*	*128.2*	*53.9*	*107.3*	*91.1*	*73.8*	*89.3*	*111.7*	*36.5*	*2,383.7*
Local	50.4	73.7	77.2	63.1	48.2	56.3	37.5	51.0	77.1	90.1	6.0	1,869.7
Foreign	23.4	39.0	65.9	73.4	10.4	56.2	58.6	26.1	17.0	29.5	39.6	745.5
Foreign (excl FDW)	*15.6*	*31.0*	*54.5*	*65.2*	*5.7*	*50.9*	*53.7*	*22.8*	*12.1*	*21.5*	*30.5*	*514.0*

Source Ministry of Manpower (2016b)

Table 4.6 Median gross monthly income from full-time employed residents aged 15 and above

Industry	2014	2015
Total	$3,770	$3,949
Manufacturing	$4,210	$4,437
Construction	**$3,480**	**$3,790**
Services	$3,750	$3,844
Others	$4,060	$4,292

Source Ministry of Manpower (2015, 2016a)

compared to other industries which rose only 5.4% (Manufacturing), 2.5% (Services) and 5.4% (Others) (Ministry of Manpower, 2015, 2016a).

In summary, the findings showed that Singapore faces a serious ageing workforce problem in the construction industry as more than 80% of the workforce are aged 40 years old and above. This proportion is unnerving because the younger workforce have not recognized the industry to be a preferred career choice. Based on the above observations, the construction industry have the lowest gross monthly income, longest weekly working paid hours and overtime hours. Furthermore, with the negative image perceived by Singaporeans, it is more unlikely that they would want to take up a job in the construction industry (Ng, 2014).

4.5.2 South Korea's Context

Based on the statistics from the Korean Statistical Information Service (KOSIS) in 2015, the construction industry constituted 7% of the entire workforce in South Korea (Korean Statistical Information Service, 2015a, 2015b). The construction industry is still one of the major industries that have contributed immensely to South Korea's gross domestic product (GDP) despite its perceptibly low make-up in the entire workforce. Table 4.7 shows that the growth rate in the construction industry has improved by 2.2% as compared to the 2014 figure. Furthermore, South Korea's overall GDP has grew by 2.6% in 2015 and the construction industry has a corresponding growth rate of 3%. This is 0.4% better than the overall GDP growth

Table 4.7 GDP and sectorial growth rates in 2013, 2014 and 2015 (in %)

GDP and sectorial	2013	2014	2015
Overall GDP growth rate	**2.9**	**3.3**	**2.6**
Agriculture, forestry and fishing	3.1	3.6	−1.5
Mining, quarry and manufacturing	3.6	3.5	1.2
Manufacturing	3.6	3.5	1.3
Electricity, gas and water supply	−0.3	2.6	6.2
Construction	**3.0**	**0.8**	**3.0**
Services	2.9	3.3	2.8
Taxes less subsidies on product	1.5	4.5	5.3

Source Bank of Korea Economic Statistics System (2016)

Table 4.8 Employed residents aged 15 years and above in the Korean construction industry between 2013 and 2015 (in thousands)

Construction industry	Total	15–29	30–39	40–49	50–59	60 and above
2013	**1,778**	102	348	613	554	161
2015	**1,824**	113	310	588	598	218

Source Korean Statistical Information Service (2015a)

rate. This growth information shows the significance of the construction industry to South Korea's economic performance.

Table 4.8 shows 1,824,000 resident workforce employed in the construction industry in 2015. By comparing the statistics between 2013 and 2015, there was an increase of 46,000 resident workforce in South Korea. However, the main bulk of the increase came from the workforce group aged 50 years and above. In contrast to the increase, there was a sharp decline of the construction workforce in the age groups of 30–39 years and 40–49 years. There were 86,000 foreigners employed in the construction industry (Statistics Korea, 2015). This means that foreigners only constituted about 5% of the entire construction workforce. This proportion was far less than the proportion in Singapore where foreigners constituted 75% of the workforce in the construction industry.

Lastly, based on the data provided by the Ministry of Employment and Labor, South Korea as shown in Appendix D, the construction industry in South Korea is one of the few industries where its wages and total working hours is above average (₩3,011,771 as compared to ₩2,926,186) and below average (181.9 h as compared to 187.9 h) respectively (Ministry of Employment and Labor, 2016). Even though the construction industry has regular working hours that were higher than the average, this was compensated by the over time working hours which were 3 times lesser than the average.

Based on the various statistical sources, the construction industry in South Korea seemed to perform better than the Singapore construction industry. South Korea's construction sector has average wages higher than the overall average wages and average total working hours lower than the overall average total working hours. Furthermore, 95% of the South Korean construction workforce consisted of local residents. Nevertheless, South Korea is facing an ageing workforce issue that is similar to Singapore. However, most of the downside in the construction industry that plagued Singapore does not seem to be applicable in the South Korean context such as those relating to better wages and shorter working hours in the latter.

4.6 Millennials

Millennials are generally those who are born after the 1980's (Knouse, 2011; Hassing, 2016; Bannon, Ford & Meltzer, 2011). Millennials are the most diversified workforce among the other two generations (Baby Boomers and Gen X) (Hassing,

2016). Moreover, the Millennials employees are more technologically well informed and highly educated (Bannon et al., 2011). As more than 50% of Millennials would be entering the workforce within the next few years, they are poised to be a powerful driver of the economy (Bannon et al., 2011). However, retaining Millennials employees poses an ultimate challenge for organizations as they thrive on new opportunities, value WLB and seek to involve in decision making (Bannon et al., 2011; Aruna & Anitha, 2015). Hence, it is important for organizations to understand the needs of the Millennials workforce in order to manage them and to utilize their potential to the fullest.

Most Millennials employees in the developed countries were raised in a sheltered environment. They also benefitted from the economic boom experienced by the country (Lee, 2016). Millennials employees are perceived to be more demanding in nature as compared to other generations (Schofield, 2014). Furthermore, they expect things to go by their way and prefer sourcing for jobs that cater to their needs and wants (Schofield, 2014). The study conducted by the Tripartite Alliance for Fair Employment Practices (TAFEP) (2010) found that the Millennials employees in Singapore value WLB, career development and good relationships with their employer. It has been said that the Millennials employees emphasize Work–Life Balance as they do not want to repeat what was perceived to be a mistake of their parents who toiled for long hours at the expense of their loved ones (Bannon et al., 2011).

As mentioned earlier, Millennials employees are those born after the 1980's. This places the oldest Millennials employee at 36-year of age in the context of this current study. However, Singapore's official statistics do not show the exact age range of the Millennials employees. Hence, the proportion method was used to extract the numbers of employees belonging to the Millennials group. For example, the age group of 35 to 39 years (spread over 5 years) have 200,000 employees. By using the proportion method, there are 80,000 employees in the Millennials group.

Table 4.9 shows that 33% of the entire Millennials group in Singapore are either working or actively seeking for a job. In addition, the Millennials group constitutes

Table 4.9 Key Characteristics of Millennials Labor Force Status (Residents) in Singapore (2015)

2015 figure (Singapore)	0–4	5–9	10–14	15–19
Resident population	183,575	204,452	214,388	242,902
Labor force	Nil	Nil	Nil	40,200
Labor participation rate	Nil	Nil	Nil	15.8%
	Nil	Nil	Nil	6,352
2015 figure (Millennials)	20–24	25–29	30–34	35–39 (35–36)
Resident population	264,127	264,127	290,619	301,067 (60,213)
Labor force	166,200	219,400	243,700	266,800 (53,360)
Labor participation rate	65.3%	90.3%	90.2%	89.1%
	108,529	198,118	219,817	237,719 (47,544)
% of Millennials entering the workforce		580,360/1,781,682 × 100% =**32.6%**		
% of Millennials in the entire workforce		580,360/2,232,300 × 100% = **26.0%**		

Source Ministry of Manpower (2016a)

Table 4.10 Key characteristic of Millennials Labor Force Status in South Korea (2016)

2016 figure (South Korea)	0–4	5–9	10–14	15–19
Resident population	2,225,116	2,359,473	2,376,960	3,183,878
Labor force	Nil	Nil	Nil	2,993,000
Labor participation rate	Nil	Nil	Nil	9.9%
	Nil	Nil	Nil	297,000
2015 figure (Millennials)	20–24	25–29	30–34	35–39 (35–36)
Resident population	3,553,038	3,179,502	3,584,698	4,002,012 (1,600,805)
Labor force	3,077,000	3,351,000	3,628,000	3,920,000 (1,568,000)
Labor participation rate	52.1%	76.5%	77.4%	76.3%
	1,603,000	2,562,000	2,808,000	2,991,000 (1,196,400)
% of Millennials entering the workforce	8,466,400/22,063,470 × 100% =**38.4%**			
% of Millennials in the entire workforce	8,466,400/27,524,000 × 100% = **30.8%**			

Source Korean Statistical Information Service (2016)

26% of the entire workforce in 2015. Table 4.10 similarly shows that 38% of the entire Millennials group in South Korea are either working or actively seeking for a job and they constituted 31% of the entire workforce in 2016. With the rising numbers of the Millennials group expecting to enter the workforce in Singapore and South Korea it is important for organizations to understand the characteristics of the Millennials workforce now to attract and retain them as well as to utilize their skills and expertise positively.

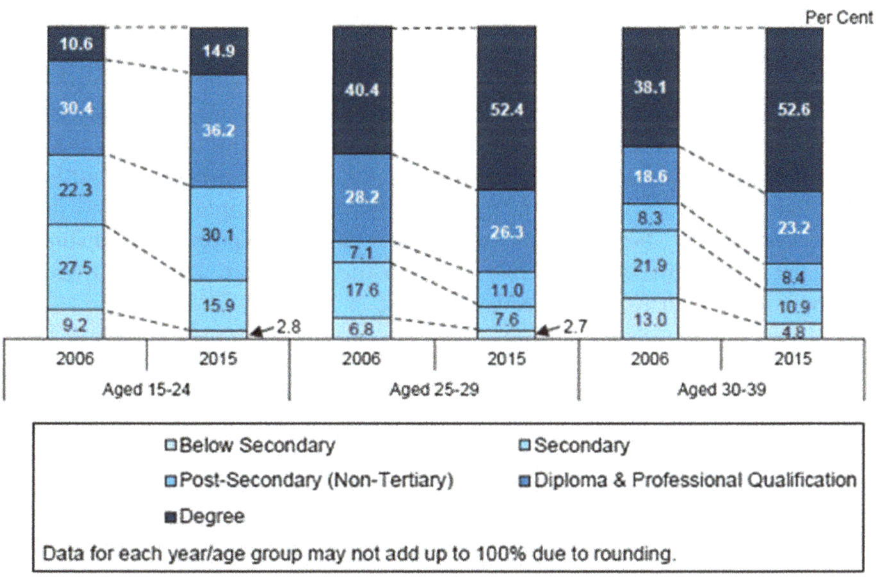

Fig. 4.2 Highest qualifications attained for resident labour force in Singapore between 2006 and 2015. *Source* Ministry of Manpower (2016a)

Table 4.11 Trends in educational attainments for Koreans aged 25–34 years in 2005 and 2015

Highest educational attainment aged 25–34 (2005 and 2015)						
Education level	Below upper secondary (s)		Upper secondary or post-sec non tertiary		Tertiary	
Year	2005	2015	2005	2015	2005	2015
Population (%)	3	2	46	29	51	69

Source Organization for Economic Co-operation and Development (2016)

By comparing the two bar charts presented in Fig. 4.2, the highest education level attained by Singaporeans aged 15–39 years has increased significantly in 2015. Similarly, Table 4.11 shows the highest education level attained by Koreans aged 25–34 years. The table also clearly shows that the percentage of Koreans attaining tertiary level or higher education is 69% as compared to 51% in 2005. With the Millennials workforce being better educated, they are able to gain access to more job choices (Seah, 2013). When more job choices are available to them, it makes organizations harder to retain this group of employees as they may choose to leave if given the opportunity. Therefore, it is critical for organizations to anticipate this problem and be one step ahead of their competitors.

4.7 Summary

It is noteworthy that even though Singapore and South Korea generally subscribe to Confucian ethos, the intensity of the traits related to Confucianism can vary for different countries as reflected by the scores in the Hofstede six cultural dimensions. These scores are critical to gain an understanding of the comparative Work–Life Balance between the two countries. In addition, Singapore and South Korea have many things in common such as those relating to the contributions of the construction industry to GDP, median gross monthly income and the proportion of Millennials employees in the labor force. However, the employment ratio in the construction industry between the two countries have more pronounced distinction where the ratio between the locals and the foreigners are 25:75 and 95:5 in Singapore and South Korea respectively.

References

Abe, M. (2006). *The developmental state and educational advance in East Asia*. Retrieved from http://www.rrojasdatabank.info/devstate/devstaasia.pdf.

Aruna, M., & Anitha, J. (2015). Employee retention enablers: Generation Y employees. *SCMS Journal of Indian Management, 12*(3), 94–103.

Bank of Korea Economic Statistics System. (2016). *Main annual indicators*. Retrieved September 29, 2016 from http://ecos.bok.or.kr/flex/EasySearch_e.jsp.

Bannon, S., Ford, K., & Meltzer, L. (2011). Understanding millennials in the workplace. *CPA Journal, 81*(11), 61–65.

Department of Statistics Singapore. (2016). *Singapore in figures 2016*. Retrieved September 12, 2016 from https://www.singstat.gov.sg/docs/default-source/default-document-library/publications/publications_and_papers/reference/sif2016.pdf.

Hagemann, R. P., & Nicoletti, G. (1989). *Ageing populations: Economic effects and implications for public finance*. Paris, France: OECD.

Hampden-Turner, C., & Trompenaars, A. (1998). *Riding the waves of culture*. New York: McGraw-Hill.

Hassing, J. (2016). Generation Y: Improving employee engagement and retention through better communication (D.H.A.). Capella University.

Hofstede, G. (2001). *Culture's consequences: Comparing values, behaviors, institutions and organizations across nations*. Thousand Oaks, Calif.: Sage Publications.

Hofstede, G., & Bond, M. (1988). The Confucius connection: From cultural roots to economic growth. *Organizational Dynamics, 16*(4), 5–21. https://doi.org/10.1016/0090-2616(88)90009-5.

Hofstede, G., Hofstede, J. G., & Minkov, M. (2010). *Cultures and organizations: Software of the mind* (3rd ed.). New York: McGraw-Hill.

Knouse, S.B. (2011). Managing generational diversity in the 21st century. *Competition Forum, 9* (2).

Korean Statistical Information Service. (2015a). *Employed persons by industry*. Retrieved September 28, 2016 from http://kosis.kr/eng/statisticsList/statisticsList_01List.jsp?vwcd=MT_ETITLEandparentId=L#SubCont.

Korean Statistical Information Service. (2015b). *Summary of economically active population by nationality*. Retrieved September 28, 2016 from http://kosis.kr/eng/statisticsList/statisticsList_01List.jsp?vwcd=MT_ETITLEandparentId=L#SubCont.

Korean Statistical Information Service. (2016). *Resident population in Five-Year age groups*. Retrieved September 20, 2016 from http://kosis.kr/eng/statisticsList/statisticsList_01List.jsp?vwcd=MT_ETITLEandparentId=L#SubCont.

Lee, K. C. (2014). Factors that influence the decisions of working adults to enter the construction industry. Unpublished undergraduate dissertation, School of Design and Environment, National University of Singapore.

Lee, Y. L. (2016). Perception of Singapore's Generation Y on built heritage conservation. Unpublished undergraduate dissertation, School of Design and Environment, National University of Singapore.

Lingard, H., & Francis, V. (2009). *Managing WLB in Construction*. Hoboken: Taylor and Francis.

Macleod, C. (2013). *Coming of age: the impacts of an ageing workforce on Australian business*. Retrieved from https://www.chandlermacleod.com/media/chandler-macleod-2013/files%20for%20content%20copying/005315-desktop-cm1278_05_13_whitepaper_coming_of_age_lr.pdf.

Ministry of Employment and Labor. (2016). *Survey on wage and working hours by regional*. Retrieved September 29, 2016 from http://laborstat.molab.go.kr:8081/OLAP/Analysis/stat_OLAP.jsp?org_id=118andtbl_id=DT_118N_M.

Ministry of Manpower. (2015). *Labour force in Singapore 2014*. Singapore: Ministry of Manpower.

Ministry of Manpower. (2016a). *Labour force in Singapore 2015*. Singapore: Ministry of Manpower.

Ministry of Manpower. (2016b). *Labour market 2015*. Singapore: Ministry of Manpower.

Ministry of Trade and Industry. (2014). *Economic survey of Singapore 2013*. Retrieved September 13, 2016 from https://www.mti.gov.sg/ResearchRoom/SiteAssets/Pages/Economic-Survey-of-Singapore-2013/FullReport_AES2013.pdf.

Ministry of Trade and Industry. (2015). *Economic survey of Singapore 2014*. Retrieved September 13, 2016 from https://www.mti.gov.sg/ResearchRoom/SiteAssets/Pages/Economic-Survey-of-Singapore-2014/FullReport_AES2014.pdf.

Ministry of Trade and Industry. (2016). *Economic survey of Singapore 2015*. Retrieved September 11, 2016 from https://www.mti.gov.sg/ResearchRoom/SiteAssets/Pages/Economic-Survey-of-Singapore-2015/FullReport_AES2015.pdf.

Ng, J. W. X. (2014). Singapore's population White Paper (2013) and implications of cultural diversities on construction productivity. Unpublished undergraduate dissertation, School of Design and Environment, National University of Singapore.

Oh, T. (1991). Understanding managerial values and behavior among the gang of four: South Korea, Taiwan, Singapore and Hong Kong. *Journal of Management Development, 10*(2), 46–56. https://doi.org/10.1108/02621719110141095.

Organization for Economic Co-operation and Development. (2016). *Average annual hours actually worked per worker*. Retrieved June 6, 2017 from https://stats.oecd.org/Index.aspx?DataSetCode=ANHRS.

Pettinger, T. (2016). *The impact of an ageing population on the economy|Economics Help*. Retrieved October 21, 2016 from http://www.economicshelp.org/blog/8950/society/impact-ageing-population-economy/.

Schofield, C. P. (2014). *What do Generation Y really want? HRZone*. Retrieved September 14, 2016 from http://www.hrzone.com/engage/employees/what-do-generation-y-really-want.

Seah, Y. L. E. (2013). *Motivating locals to contribute to Singapore construction industry*. Unpublished undergraduate dissertation, School of Design and Environment, National University of Singapore.

Singapore—Geert Hofstede. Geert-hofstede.com. Retrieved September 8, 2016 from https://geert-hofstede.com/singapore.html.

South Korea—Geert Hofstede. Geert-hofstede.com. Retrieved September 8, 2016 from https://geert-hofstede.com/south-korea.html.

Statistics Korea. (2015). *2015 foreigner labour force survey*. South Korea: Statistics Korea.

Tripartite Alliance for Fair Employment Practices. (2010). *Harnessing the potential of Singapore's multi-generational workforce*. Retrieved September 14, 2016 from https://www.tafep.sg/sites/default/files/Publication%20-%20Harnessing%20the%20Potential%20of%20Singapore%27s%20Multi-Generational%20Workforce.pdf.

Wells, J. (2001). *The construction industry in the twenty-first century: Its image, employment prospects and skill requirements*. Geneva: International Labour Organization.

Chapter 5
Research Design and Methodology

Abstract This chapter discusses the formulation of the research design and adoption of the appropriate research method. This includes the formulation of the survey questionnaire, the method of analyzing the data using descriptive and inferential statistics as well as providing qualitative inputs from the interview findings and background information of the interviewees.

5.1 Overview of Chapter

This chapter presents the research design and methodology adopted for the study. It comprises three parts: research design, data collection methods and data analysis methodology. The research design outlines the overall strategy adopted to address the research aims and objectives. The data collection methods outline the set of qualitative and quantitative methods adopted to collate the empirical data needed to analyze the conceptual framework. Finally, the data analysis methodology discusses the findings of the survey results.

5.2 Research Design

Figure 5.1 shows the process for the research design. The research design in this study can be categorized into three stages: identification, analysis and evaluation. The first stage of this research is the identification stage. It starts by establishing the research problems, aim, objectives and scope. It is important to reiterate that the aim of this research is to operationalize a Work–Life Balance (WLB) framework that can be used by researchers universally and to determine if the path in achieving Work–Life Balance differs in Singapore and South Korea. After identifying the research agenda, an extensive literature review was carried out to establish a foundation for the research objectives as well as lay the groundwork for the conceptual framework. The literature review covers the following three areas:

L. Sui Pheng and B. K. Q. Chua, *Work–Life Balance in Construction*,
Management in the Built Environment,
https://doi.org/10.1007/978-981-13-1918-1_5

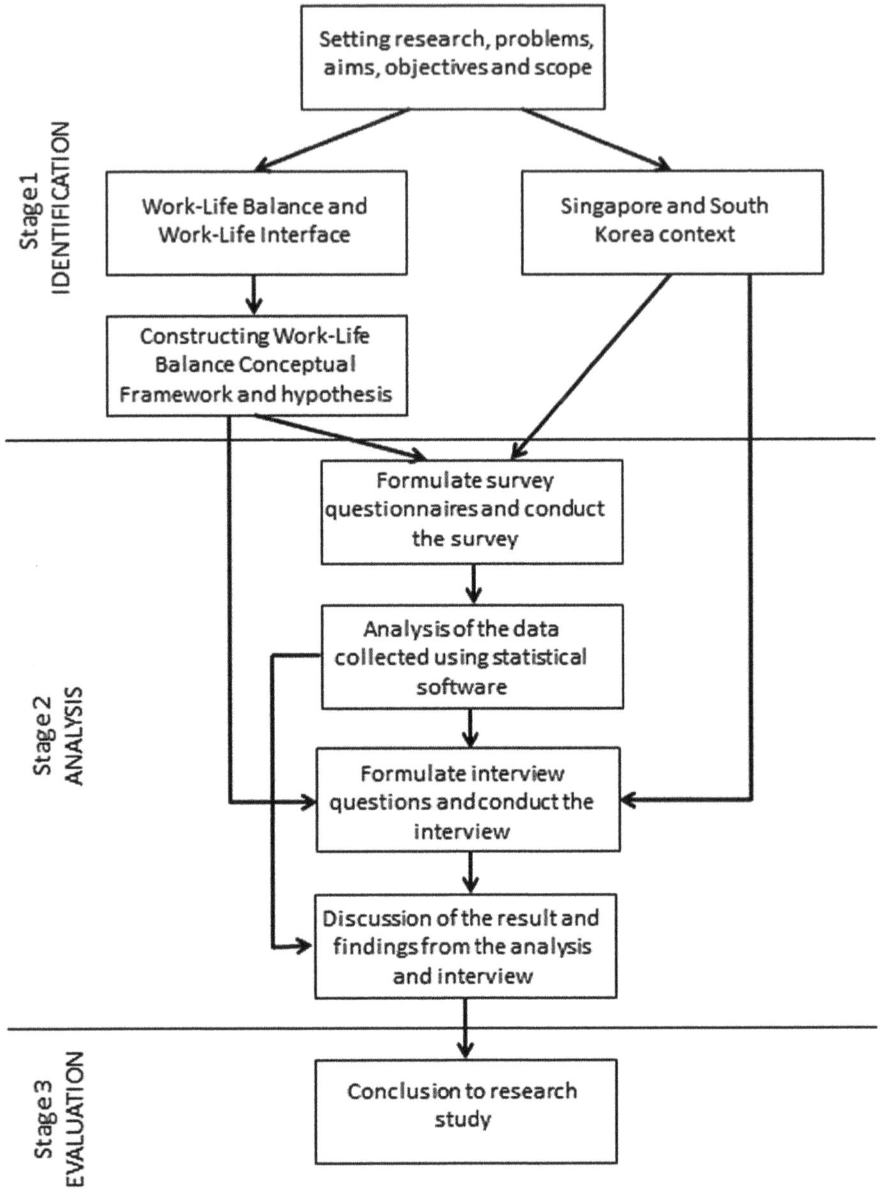

Fig. 5.1 Research design and methodology

- Identify past WLB research work and its associated Work–Life Interfaces.
- Formulate a conceptual framework for WLB to assist in formulating a set of hypotheses.

- Identify comprehensive studies from the research scope focusing on the Singapore and South Korean construction industry and the Millennials workforce.

The second stage of the research is the analysis stage. Both the quantitative and qualitative methods were used to conduct the research analysis. The quantitative method was used to analyze the conceptual framework while the qualitative method was adopted to validate the result of the findings. Following the development of the conceptual framework, the quantitative approach was conducted through a survey where questionnaires were designed based on the indicators presented in the conceptual framework. Each of the questions was specially formulated to meet the purpose of this research. The questionnaire survey was specifically targeted at Singaporean and South Korean Millennials employees working in the construction industry; namely with main contractor firms in general. After conducting the questionnaire survey, the results were collated and analyzed using statistical software. Descriptive and inferential statistical analyses were conducted on the survey data to identify pertinent findings.

Descriptive statistical analysis was conducted to identify the mean scores of the survey results from both Singapore and South Korea. Inferential statistical analysis was conducted to test the hypotheses for this research as well as to establish the validity and reliability of the measurement. After the completion of the inferential statistical analysis, the qualitative data collection method was conducted using interviews to validate the results obtained from the empirical study as well as to gain in-depth information and insights on how organizations implement Work-Life Balance. Interview questions were formulated based on the conceptual framework, the research hypotheses and the results of the statistical analysis. In the final part, findings from the statistical analysis results and interviews were discussed. The last stage of the research, the evaluation stage, concludes this study by discussing the theoretical and practical implications, strengths and limitations of this research as well as recommendations for further study.

5.3 Data Collection Methods

5.3.1 Survey Questionnaire

The questionnaire is an ideal platform to engage with the respondents as this allows researchers to assess their attitudes and behaviors quantitatively, usually in the form of a Likert scale. The questionnaire was designed for the purpose of analyzing the conceptual framework designed for this study. As presented earlier, Table 3.1 shows the overall summary of the indicators and sub-indicators for this research. A series of questions focusing on each indicator was adopted whereby the respondents answer the questions based on a five-point Likert scale. To cater for two different nationalities, the survey was crafted in two languages: English and

Korean as shown in Appendix E and Appendix F respectively. The surveys were conducted during November–December 2016 and February–May 2017 in South Korea and Singapore respectively.

5.3.2 *Survey Design*

All measures in this research uses a five-point Likert scale where 1 = Strongly Disagree, 3 = Neither Agree nor Disagree and 5 = Strongly Agree. The rating scales allow the results to be analyzed and interpreted statistically. In total, there were 111 questions separated into two categories: Work and Life Domains. Most of the questions were identified from past research works based on the literature review, while some were designed specifically for this study to suit the purpose of this research. The details of each section in the questionnaire are discussed below. The survey questionnaire is divided into three parts: Work–Life Balance Assessment, Work–Life Balance Indicator and Work–Life Balance Outcome. The Work–Life Balance Assessment part consists of questions pertaining to the Work–Life Interfaces which were used to measure the respondent's level of WLB. The Work–Life Balance Indicator part includes questions that directly seek the respondent's views on their level of WLB. Finally, the Work–Life Balance Outcome part is used to find out whether the achievement of WLB by respondents has a positive outcome towards their work and life domain.

The questionnaire is divided into eight sections; six of which belonged to Work–Life Balance Assessment, one belonged to Work–Life Balance Indicator and one belonged to Work–Life Balance Outcome. The design of the questionnaires for each indicator and its sub-indicators are discussed below.

5.3.2.1 Work-Life Balance Assessment

i. Work Demands/Life Demands

This study used a 6-item scale on each domain to measure Work Demands and Life Demands. Each sub-indicator consists two items identified from past research work. As shown in Table 5.1, role conflict, role ambiguity and involvement were used interchangeably in work and life domains. Thus, items associated with each sub-indicator might also contain similar question, provided it is able to fit into both domains. Details of each sub-indicator are explained below:

Table 5.1 Work and life demands' sub indicators

	Work domain	Life domain
Sub-indicators	Work role conflict	Life role conflict
	Work role ambiguity	Life role ambiguity
	Work involvement	Life involvement

- Work and Life Role Conflict

A 2-item work role conflict scale and a 2-item life role conflict scale were developed by Rizzo, House, and Lirtzman (1970). This was used to measure the conflict an individual faced when performing the particular role. One example of an item from the work domain is "I work under incompatible policies and guidelines". One example of an item from the life domain is "I have many roles in life that operate quite differently from one another".

- Work and Life Role Ambiguity

A 2-item work role ambiguity scale and a 2-item life role ambiguity scale were developed by Rizzo et al. (1970). This measures the level of ambiguity an individual faced with regards to his/her objectives, duties and responsibilities when performing the particular role. One example of an item from the work domain is "I do know exactly what my responsibilities are at work". One example of an item from the life domain is "I do not know exactly what is expected of me in life".

- Work and Life Involvement

A 2-item work involvement and a 2-item life involvement scale were developed by Quinn and Staines (1979). This measures the level of psychological involvement of a person to his/her particular role. One example of an item from the work domain is "I would like to spend more time in working". One example of an item from the life domain is "The most important things that happen to me involve my life".

A summary of the measures for Work and Life Demands' sub-indicators based on the literature review is presented in Table 5.2.

ii. Boundary Spanning Demands

Items associated with the Boundary Spanning Demands shown in Table 5.3 were designed specifically for this current study as there were no available empirical studies prior to this research. This study has developed an 8-item scale where each sub-indicator (four in total) contains two items. The details of each sub-indicator are explained below:

Table 5.2 Summary of measures for work and life demands' sub-indicators

Sub-indicators	No. of item(s)	Authors
Work demands		
Work role conflict	2	Rizzo, House, and Lirtzman (1970)
Work role ambiguity	2	Rizzo, House, and Lirtzman (1970)
Work involvement	2	Quinn and Staines (1979)
Life demands		
Life role conflict	2	Rizzo, House, and Lirtzman (1970)
Life role ambiguity	2	Rizzo, House, and Lirtzman (1970)
Life involvement	2	Quinn and Staines (1979)

Table 5.3 Boundary spanning demands' sub indicators

	Work domain	Life domain
Sub-indicators	Work overload	Life activities at work
	Bring work to home	Life interruptions and distraction

(a) Work Domain

- Work Overload

The concept of work overload came about due to the nature of the construction industry. As mentioned by Lee (2014), the construction industry has caused stress build-up due to several reasons including huge workloads and/or long working hours which affect both work and life of an individual. Hence, work overload was added to find out the impact ithas on an individual. One example of such an item is "My job has often overloaded me with works that must be completed within a tight deadline".

- Bring Work to Home

The concept of bringing work to home was based on Voydanoff's (2005) study where the term "bring work home" was coined. This means an individual is spilling over the work domain into his/her life domain, affecting two domains at the same time. An example of such an item is "Bringing work to home has negatively affected my personal life".

(b) Life Domain

- Life Activities at Work

The concept of life activities at work came about due to the rise of information technology (IT) influence in everyday life. As observed by Valcour and Batt (2003), IT has allowed life activities to be brought to work through the use of communication tools like cellphones or electronic mails. Hence, if the organization is flexible in allowing employee to use their communication tools to communicate with friends or family, it will allow life activities to be spilled over to work (Valcour and Batt 2003). An example of such an item is "My job allows me to bring life activities to work (e.g., communicating with friends and family on the phone)".

- Life Interruptions and Distraction

The concept of life interruptions and distraction was identified from Voydanoff's (2005b) study where it was explained that performing life activities at work can ease the transitions across domains and creating interruptions and distraction to both domains. An example of such an item is "I often allow my personal life issues to be brought over to work".

A summary of the measures for Boundary Spanning Demands' sub-indicators based on a review of the literature and as identified by the research team of this current study is shown in Table 5.4.

Table 5.4 Summary of measures for boundary spanning demands' sub-indicators

Sub-indicators	No. of item(s)	Authors
Boundary spanning demands (work domain)		
Work overload	2	Research team
Bringing work to home	2	Research team
Boundary spanning demands (life domain)		
Life activities at work	2	Research team
Life interruptions and distraction	2	Research team

iii. Work Resources/Life Resources

This current study used a 6-item scale in each domain to measure Work and Life Resources. These six items are divided into three sub-indicators as shown in Table 5.5 consisting of two items each. Carlson, Kacmar, Wayne, and Grzywacz (2006) used job salience, life salience, developmental experiences and autonomy as antecedents of Work–Life Enrichment. Life Harmony and life rewards on the other hand were developed for this study (based on other relevant literature review) to fit the purpose of this study. The details of each sub-indicator are explained below:

(a) Work Domain

- Job Salience

This study has adopted the 2-item job salience scale developed by Lodahl and Kejner (1965). This scale was used to measure the degree to which work is central to an individual's life. An example of such an item is "I am very much involved personally in my work".

- Developmental Experience

This study has adopted the 2-item developmental scale developed by Wayne, Shore, and Liden (1997). This scale was used to measure the degree to which developmental opportunities were provided by organization to the individuals. An example of such an item is "At the position that I have held at this organization, I have often been assigned projects that enabled me to develop and strengthen new skills".

- Autonomy

This study has adopted the 2-item autonomy scale developed by Spreitzer (1995). This scale measures the degree of freedom an individual has while doing his/her

Table 5.5 Work and life resources' sub indicators

	Work domain	Life domain
Sub-indicators	Job salience	Life salience
	Developmental experiences	Life harmony
	Autonomy	Life (psychological) rewards

job. An example of such an item is "I have considerable opportunity for independence and freedom in how I do my job".

(b) Life Domain

- Life Salience

Life salience is adapted from job salience based on Carlson et al.'s (2006) study. The scale used for life salience was adapted from Lodahl and Kejner's (1965) survey. An example of such an item is "Most things in my life are more important than work".

- Life Harmony

The concept of life harmony was based on Mcmillan, Morris, and Atchley (2010) study, where harmony was deemed as successful when individuals are able to achieve a sense of fulfillment while balancing their life responsibilities. The 2-item scale was designed to measure the level of life harmony an individual has when enacting different life roles. An example of such an item is "My duty as a friend, family member or other non-work role is co-operating with each other harmoniously".

- Life Rewards

This study has designed a 2-item life rewards scale to measure the level of psychological rewards received by individuals. The concept of life rewards was based on Voydanoff's (2005b) study that used psychological rewards as a resource that would help improve the performance of individuals when they are applied across domains. Psychological rewards include compliments and recognitions. An example of such an item is "I am satisfied with the recognition that I gotten from my families and friends".

A summary of the measures for Work and Life Resources' sub-indicators based on the literature review is shown in Table 5.6.

iv. Boundary Spanning Resources

Items associated with Boundary Spanning Resources shown in Table 5.7 were designed for the purpose of this study as there were no available empirical studies

Table 5.6 Summary of measures for work and life resources' sub-indicators

Sub-indicators	No. of item(s)	Authors
Work resources		
Job salience	2	Lodahl and Kejner (1965)
Developmental experience	2	Wayne, Shore, and Liden (1997)
Autonomy	2	Spreitzer (1995)
Life resources		
Life salience	2	Lodahl and Kejner (1965)
Life harmony	2	Research team
Life rewards	2	Research team

Table 5.7 Boundary spanning resources' sub-indicators

	Work domain	Life domain
Sub-indicators	Flexible work schedule	Participate in workshops for personal development
	Time-off for non-work purposes	Assign time for work-purposes

prior to this current research. This study has created an 8-item boundary spanning resources scale where each sub-indicator (total of fou) contains two items. Details of each sub-indicator are explained below:

(a) Work Domain

- Flexible Work Schedule

The concept of flexible work schedule was based on Voydanoff's (2005b) study who described it as organization policies that help employees accommodate their family responsibilities without reducing the amount of work done. An example of such an item is "I am given flexible work schedule whereby I can work at a time that best suits me".

- Time-Off for Non-work Purposes

The concept of time-off for non-work purposes was based on Voydanoff's (2005b) time off for family concept, and who described it as the family-support policy that allows employees to reduce their work time to accommodate their family responsibilities with no career penalties at stake. However, this present study has replaced time-off for family to time-off for non-work purposes as the latter covers a wider perspective. An example of such an item is "My organization allowed me to apply time-off for non-work purposes".

(b) Life Domain

- Participate in Workshops for Personal Development

The concept of "participate in workshops for personal development" was based on Gadd's (2012) work, who mentioned that personal development allows individuals to enhance their skills and knowledge that can be used in both their career and life (Gadd, 2012). By participating in such activities, individuals are exposed to the external environment where they can understand themselves better and to uncover their hidden potentials (Gadd, 2012). An example of such an item is "I often participate in workshops for personal development".

- Assign Time for Work Purposes

The concept of assigned time for work purposes was adapted from time-off for non-work purposes. This present study views that the concept works in both ways from work to life and from life to work. Hence, assign time for work purposes

Table 5.8 Summary of measures for boundary spanning resources' sub-indicators

Sub-indicators	No. of item(s)	Authors
Boundary spanning resources (work domain)		
Flexible work schedule	2	Research team
Time-off for non work purposes	2	research team
Boundary spanning resources (life domain)		
Participate in workshops for personal development	2	Research team
Assign time for work-purposes	2	Research team

posits that individuals would reduce their non-work time to commit to their work responsibilities with no impact to their own personal life. An example of such an item is "I will assign time during my free days for work purposes".

A summary of the measures for Boundary Spanning Resources' sub-indicators based on a review of the literature is shown in Table 5.8.

v. Proposed Measures

Items associated with Proposed Measures shown in Table 5.9 were designed for the purpose of this present study because there were no available empirical studies prior to this research. The Proposed Measures were created in such a way that these can relate to the nature of the construction industry and the targeted respondents. This present study has developed a 12-item boundary scale where each sub-indicator (total of twelve) contains one item. The details of each sub-indicator are explained below:

(a) Work Domain

- Take Less Demanding Task

The concept of take less demanding task was identified from Voydanoff's (2005b) study. This occurs when an employee has too much work on hand which leads to high work demand. Voydanoff (2005b) suggested that this measure can help individuals reduce or eliminate any incompatibility between work and life demands. An example of such an item is "I prefer taking up less demanding task than the work I am dealing right now".

Table 5.9 Proposed measures' sub indicators

Reduce demands	Work domain	Life domain
Sub-indicators	Take less demanding task	Reduce time with love ones
	No work after 6 pm	Limit usual lifestyles
	Organization plays A part	Reduce role responsibility
Increase resources	Work domain	Life domain
Sub-indicators	Assign more challenging task	Take up new sports/hobby
	More break interval	Plan weekly activity beforehand
	Provide more freedom	Assign more personal time

- No Work After 6 pm

This concept was designed to fit the nature of the construction industry. As observed by Lee (2014), long working hours are inevitable in the construction industry due to tight deadlines and heavy workload. Furthermore, as shown in Appendix B, the construction industry in Singapore has the longest weekly overtime paid hours as compared to other industries. Hence, no work after 6 pm is used to measure the degree to which an individual would like to shut off from work after official working hours. An example of such an item is "I hope to shut off from work after 6 pm".

- Organization Plays a Part

The concept of organization plays a part was based on Hall and Richter's (1988) work. They mentioned that organizations could influence both work and life domain of an individual as policies could be set out to effectively control the boundary or transition between work and life (Hall and Richter, 1988). Hence, it is essential for organizations to help look for ways to reduce or manage conflicts between an employee's private and working life (Hall and Richter, 1988). An example of such an item is "I hope my organization can play a bigger role in helping me cope with the high demand at work".

- Assigning More Challenging Task

The concept of assigning more challenging task was based on Lingard and Francis' (2009) work. They noted that employees who are pursuing promotion in the organization are more interested in receiving more challenging tasks to support their career advancement (Lingard & Francis, 2009). Hence, this concept could be viewed as an extrinsic reward. An example of such an item is "I want to have more challenging tasks assign to me".

- More Break Interval

The concept of more break interval was also based on Lingard and Francis' (2009) study. It was suggested that individuals be given longer breaks during time of inactivity without loss of pay. Such a measure can help to improve employee's efficiency and productivity in work. An example of such an item is "I hope to have more breaks in between my working hours to allow me to relax my mind".

- Provide More Freedom

The concept of provide more freedom was developed based on Lingard and Francis's (2009) work. Freedom in this case refers to the amount of flexibility that the individual has to fulfill his/her work. Lingard and Francis (2009) opined that flexibility could be associated with work practices, work arrangement or work schedule. Therefore, setting an organization policy that focuses on flexibility would enable an individual to have the freedom to produce quality work. An example of such an item is "Having more freedom in my work allows me to produce better result".

(b) Life Domain

- Reduce Time with Loved Ones

Reduce time with loved ones means an individual can reduce the demand to compensate for the work domain. Lingard and Francis (2009) deemed that individuals working in the construction industry should understand its nature such as a culture of long working hours and tight schedules. They also opined that individuals who refused to prioritize their work over loved ones were perceived as lacking commitment and deemed to be incompatible for the job. Hence, this indicator was used to assess whether prioritizing work over loved ones would allow individuals achieve better results. An example of such an item is "Reduce time with my loved ones will allow me to better focus on my work".

- Limit Usual Lifestyles

The concept of limit usual lifestyles was adapted from "taking less demanding task". This present study posited that individuals were overwhelmed by their current lifestyles that could directly and indirectly affect their life and work domain respectively. This concept was used to assess the extent to which the measure can help an individual reduce or eliminate any incompatibility between work and life demands. An example of such an item is "Limit my usual lifestyle will allow me to have better focus on my work".

- Reduce Role Responsibilities

The concept of reduce role responsibilities was developed based on Lingard and Francis's (2009) work. They found that although individual's satisfaction and fulfilment were formed as a result of successful involvement of multiple life roles, juggling with too many roles was more likely to cause conflicts between them (Lingard & Francis, 2009). Hence, it was suggested that reducing role responsibilities could allow an individual to manage roles more easily. An example of such an item is "Reduce responsibility in my non-work life will help me attain balance between my work and life".

- Take Up New Sports/Hobby

This concept was based on Lingard and Francis's (2009) work. They mentioned that leisure activities were important as these helped to provide time for individuals to recuperate from work related stresses (Lingard & Francis, 2009). Leisure in this case refers to participation in activities such as sports and taking up a new hobby. This concept of take up new sports/hobby proposed that individuals who took up new leisure activities would improve their life resources while reducing work demand. An example of such an item is "I wish to take up new sports and hobbies to switch my mind off work".

- Plan Weekly Activity Beforehand

Planning activities ahead of time allows individuals to anticipate their upcoming events (Lingard & Francis, 2009). Such activities could also indirectly motivate individuals at work as they have something to look forward to at the end of the day. In this case, the study proposed that individuals should have a weekly plan for their social life and viewed that preparing this plan was manageable to most individuals. An example of such an item is "Planning weekly activities with friends and family will allow me to have something to look forward to".

- Assign More Personal Time

Over commitment towards work and life by individuals were linked to clinical depression, decreased satisfaction with their life and burnout (Lingard & Francis, 2009). It is important for individuals to have more personal time during their non-working days to recuperate from such commitments so as to reduce the negative effects. An example of such an item is "I want to have more uninterrupted personal time during my free time to keep away from work and life".

A summary of all the measures for the Proposed Measures' sub-indicators based on a review of the literature is shown in Table 5.10.

vi. Work–Life Conflict

This study has adopted the 18-item Work–Life Conflict scale developed by Carlson, Kacmar, and Williams (2000). Of these, nine items represent the Work Domain while the other nine items represent the Life Domain. As presented in Table 5.11, these nine items were sub-divided into three parts: time-based, strain-based and behavior-based, with each containing three questions. An example from the Work Domain (time-based) is "I have to miss my life activities due to the amount of time I must spend on work responsibilities". Another example from the Life Domain

Table 5.10 Summary of measures for proposed measures' sub-indicators

Sub-indicators	No. of item(s)	Authors
Reduce demands (work domain)		
Take less demanding task	1	Research team
No work after 6 pm	1	Research team
Organization plays A part	1	Research team
Reduce demands (life domain)		
Reduce time with loved ones	1	Research team
Limit usual lifestyles	1	Research team
Reduce role responsibilities	1	Research team
Increase resources (work domain)		
Assign more challenging task	1	Research team
More break interval	1	Research team
Provide more freedom	1	Research team
Increase resources (life domain)		
Take up new sports/hobby	1	Research team
Plan weekly activity beforehand	1	Research team
Assign more personal time	1	Research team

Table 5.11 Work–life conflict's sub indicators

	Work domain	Life domain
Sub-indicators	Time-based	Time-based
	Strain-based	Strain-based
	Behavior-based	Behavior-based

(time-based) is "the time I spend with my life often cause me not to spend time in activities at work that could be helpful to my career". A summary of the measures for Work–Life Conflict's sub-indicators based on the literature review is shown in Table 5.12.

vii. Work–Life Enrichment

This study has adopted the 18-item Work–Life Enrichment scale developed by Carlson et al. (2006). Of these, nine items represent the Work Domain while the other nine items represent the Life Domain. These nine items were further sub-divided into three parts as shown in Table 5.13 and with each sub-indicator containing three questions. An example item from the Work Domain (development) is "My involvement in my work helps me understand different viewpoints and this helps me be a better person in life". Another example item from the Life Domain (affect) is "My involvement in my life puts me in a good mood and this helps me be a better worker". A summary of the measures for the Work–Life Enrichment's sub-indicators based on the literature review is shown in Table 5.14.

viii. Coping Strategies

The concept of coping strategies was developed base on Haddon and Hede's (2010) research where they identified different strategies that individuals used to cope with

Table 5.12 Summary of measures for work–life conflict's sub-indicators

Sub-indicators	No. of item(s)	Authors
Work–life conflict (work domain)		
Time-based	3	Carlson et al. (2000)
Strain-based	3	Carlson et al. (2000)
Behavior-based	3	Carlson et al. (2000)
Work–life conflict (life domain)		
Time-based	3	Carlson et al. (2000)
Strain-based	3	Carlson et al. (2000)
Behavior-based	3	Carlson et al. (2000)

Table 5.13 Work–life enrichment's sub-indicators

	Work domain	Life domain
Sub-indicators	Development	Development
	Affect	Affect
	Capital	Efficiency

Table 5.14 Summary of measures for work–life enrichment's sub-indicators

Sub-indicators	No. of item(s)	Authors
Work–life enrichment (work domain)		
Development	3	Carlson et al. (2006)
Affect	3	Carlson et al. (2006)
Capital	3	Carlson et al. (2006)
Work–life enrichment (life domain)		
Development	3	Carlson et al. (2006)
Affect	3	Carlson et al. (2006)
Efficiency	3	Carlson et al. (2006)

their work and life domain. This study has identified five of their strategies and adopted these as the sub-indicators for this research as shown in Table 5.15. Each strategy was then assigned with a 1-item scale designed for this study. Items in each sub-indicator were designed to measure the degree to which the coping strategy would help in reducing the stress (WLC) and in improving the performance (WLE) of an individual.

(a) Work and Life Domain

- Colleagues and Employer Support

Results from Haddon and Hede's (2010) research emphasized the importance of an environment that is supportive towards WLB. An example of this item is "Support from my colleagues and employer have helped me better cope with stress and improve my performance in work and life roles".

- Families and Friends Support

This indicator follows the same concept as "colleagues and employer support" but instead of the work environment, supportive life environment was introduced. An example of this item is "Support from my friends and families have helped me better cope with stress and improve my performance in work and life roles".

- Cognitive Management

Cognitive management is crucial in helping individuals actively re-appraise their situations (work or life domain) in line with their values and beliefs. If they manage well, it can help them deal with higher amount of stress or motivate them to perform better even under an environment of high role conflict (Haddon & Hede, 2010). An example of this item is "I am used to the usual work and life habits that make me have the ability to deal with high stress or low performance at work or life".

Table 5.15 Coping strategies' sub indicators

	Work and life domain	
Sub-indicators	Colleagues and employer support	Families and friends support
	Cognitive management	
	Behavioral management	
	Leisure activities	

- Behavioral Management

Haddon and Hede (2010) opined that individuals would engage in specific behavioral strategies as a form of intervention or enhancement aimed at achieving WLB. There are many behavioral types that individuals may use. Hence, in this case behavioral management is any behaviors that are deemed appropriate by an individual in certain situations. An example of such an item is "I will engage in specific behavior which is appropriate to deal with high stress or low performance at work or life".

- Leisure Activities

Leisure is another effective coping strategy that helps individuals maintain or improve their health and wellness (Haddon & Hede, 2010). There are many leisure types that individuals can participate in. These include active (e.g. playing sports, exercise) or passive (e.g. playing computer games or watching television) activities. An example of such an item is "My usual leisure activities have helped me better cope with stress and improve my performance in work and life roles".

A summary of the measures for Coping Strategies' sub-indicators based on a review of the literature is shown in Table 5.16.

5.3.2.2 Work–Life Balance Indicator

i. Work–Life Balance

This present study has adopted the 6-item Work–Life Balance scale developed by Carlson, Grzywacz, and Zivnuska (2009). This scale helps to measure the extent to which individuals are able to balance their role related expectations in both work and life domains and to tap on the social factors related perspective as well. An example of this item is "People who are close to me would say that I do a good job of balancing work and life". A summary of the measures for Work–Life Balance based on the literature review is shown in Table 5.17.

Table 5.16 Summary of measures for coping strategies' sub-indicators

Sub-indicators	No. of item(s)	Authors
Coping strategies		
Colleagues and employer support	1	Research team
Families and friends support	1	Research team
Cognitive management	1	Research team
Behavioral management	1	Research team
Leisure activities	1	Research team

Table 5.17 Summary of measures for work–life balance

Indicator	No. of item(s)	Authors
Work–life balance		
Work–life balance	6	Carlson, Grzywacz, and Zivnuska (2009)

5.3.2.3 Work–Life Balance Outcomes

i. Work Domain Outcomes/Life Domain Outcomes

Six sub-indicators were identified to explain the work and life domain outcomes of an individual. Table 5.18 shows the list of six sub-indicators. Each of these contains two items identified from past research work. The breakdown of each of the sub-indicators is explained below:

(a) Work Domain

- Job Satisfaction

This study has adopted the 2-item job satisfaction scale developed by Cammann, Fichman, and Williams (1979). This scale is used to measure the degree to which the person is satisfied with his/her job. An example for this item is “All in all, I am satisfied with my job”.

- Organizational Commitment

This study has adopted the 2-item organizational commitment scale developed by Balfour and Wechsler (1996). This scale is used to measure an individual’s commitment to the organization. An example for this item is “I am proud to be able to tell people who it is I work for”.

- Intention to Turnover

This study has adopted the 2-item intention to turnover scale developed by Seashore, Lawler, Mirvis, and Cammann (1982). This scale is used to measure the degree to which individuals have an intention to turnover. An example of this item is “I will not give up on this job easily”.

(b) Life Domain

- Family Satisfaction

This study has adopted the 2-item family satisfaction scale developed by Zabriskie and Ward (2013). This scale is used to measure the satisfaction of individuals with

Table 5.18 Work and life outcomes’ sub-indicators

	Work domain	Life domain
Sub-indicators	Job satisfaction	Life satisfaction
	Organizational commitment	Family satisfaction
	Reduce turnover intention	Psychological well-being

Table 5.19 Summary of measures for work–life balance outcomes' sub-indicators

Sub-indicators	No. of item(s)	Authors
Outcomes (work domain)		
Job satisfaction	2	Cammann, Fichman, and Williams (1979)
Organization commitment	2	Balfour and Wechsler (1996)
Reduce turnover intention	2	Seashore, Lawler, Mirvis, and Cammann (1982)
Outcomes (life domain)		
Life satisfaction	2	Zabriskie and Ward (2013)
Family satisfaction	2	Diener, Emmons, Larsen, and Griffin (1985)
Psychological well-being	2	Ryff (1989)

their family life. An example of this item is "The conditions of my family life are excellent".

- Life Satisfaction

This study has adopted the 2-item life satisfaction scale developed by Diener, Emmons, Larsen, and Griffin (1985). This scale is used to measure an individual's perception of his/her satisfaction in life. An example of this item is "In most ways my life is close to ideal".

- Psychological Well-Being

This study has adopted the 2-item psychological well-being scale developed by Ryff (1989). This scale is used to measure the outcome of an individual's perception of his/her overall well-being in life. An example of this item is "For the most part, I am proud of whom I am and the life I lead".

A summary of the measures for Work–Life Balance Outcomes' sub-indicators based on the literature review is shown in Table 5.19.

5.3.3 *Interviews*

Interview is a useful triangulation technique that helps to validate the results of an analysis by cross verifying the same information with interviewees. This serves to increase the credibility of the findings. Interviews were conducted in this present study to gain first-hand knowledge from selected interviewees relating to their views on the results of the analysis as well as providing additional insights of their current work and life situations. One Singaporean and one South Korean Millennials individual working in the construction industry (with main contracting firms) were invited to participate in the interviews and to comment on the various indicators and the significance these have on Work–Life Balance. Prior to conducting these interviews, the research team has already completed the inferential statistical analysis and the results of the findings were presented to the interviewees.

Table 5.20 Profile of interviewees

Name/Age	Job scope	Background	Verbatim report
Miss A/24	Quantity surveyor	Miss A is a Singaporean who has worked with a Singaporean main contractor for one year. She was stationed in the main office and was working on three tender projects at the time of the interview. She belonged to a middle-income family	Appendix G
Manager B/65	M&E Contracts Manager	Manager B has worked with a Grade A1 main contractor (with unlimited tendering limits) in Singapore for eleven years. He has six Millennials employees working under him at the time of the interview	Appendix H
Mr C/36	Site Engineer	Mr C is a South Korean who has worked with a South Korean main contractor for ten years. At the time of the interview, he was deployed to Singapore for two years as a senior site engineer for a project in the eastern part of Singapore	Appendix I
Director D/47	Project Director	Director D has worked with a South Korean main contractor for six years. At the time of the interview, he was a project director in a project in the eastern part of Singapore. He has worked with many South Koreans over the past few years	Appendix J

A series of questions was also asked to gain insights of their respective current work and life environment in order to highlight any information that requires special attention. Other than the Millennials participants, the research team has also invited the managers who are dealing with the Millennials employees in their respective workplace to participate in the interviews. A series of questions were also asked to find out their views on the statistical results as well as to gain insights of how the managers assist the Millennials employees to attain WLB. The profiles of the interviewees are shown in Table 5.20.

5.4 Data Analysis Methodology

The data analysis methodology analyses and discusses the findings of the survey using appropriate software. Descriptive and inferential statistical analyses were conducted using the survey returns to evaluate information from the data. The data collected was analyzed using two different software: IBM Statistical Package for the Social Sciences (SPSS) and Smart Partial Least Squares 3 (SmartPLS). In this study, the SPSS software was used to compute the mean scores of all the survey items and the results were used for presenting the descriptive statistics. The SmartPLS 3 software was used to test the conceptual framework that this study has proposed. Before computing any inferential statistics using the software, it was critical to tabulate the survey results to make it the sole platform to retrieve data in order to reduce any errors when importing to other software. In this case, Microsoft Excel was used to compile all the survey inputs.

5.4.1 Descriptive Statistics

The ranked mean scores from both the Singaporean and South Korean respondents were used to compare the different Work–Life Balance (WLB) questions from the survey results. The measurement for WLB was assessed based on three different categories: WLB Assessment, WLB indicators and WLB outcomes. In the descriptive statistics, a 4-step approach as illustrated in Fig. 5.2 was applied to the analysis to ensure that the comparisons covered are comprehensive.

Step 1: Overall Comparison—A comparison was made by ranking the mean scores of both Singapore and South Korea. Both results were compared separately to identify which indicators (from each country) have the highest and lowest mean scores. Items that were ranked in the top 5 and bottom 5 were evaluated to identify the possible reasons for the phenomena observed.

Step 2: WLB Assessment—The WLB assessment was essentially concerned with indicators used to explain the theory of WLB. There were ten indicators in this category: Work Demands, Boundary Spanning Demands, Life Demands, Life Resources, Boundary Spanning Resources, Life Resources, Proposed Measures, Work–Life Conflict, Work–Life Enrichment and Coping Strategies. The mean difference of the scores between Singapore and South Korea for each item was ranked accordingly. There were a total of 93 items and only the top 5 and bottom 5 scores were evaluated.

Step 3: WLB Indicators—The WLB indicators were items directly used to measure the level of WLB of an individual. Similar to Step 2, the mean difference in the scores between Singapore and South Korea for each similar item was ranked. There were a total of six items in this category and all the items were evaluated.

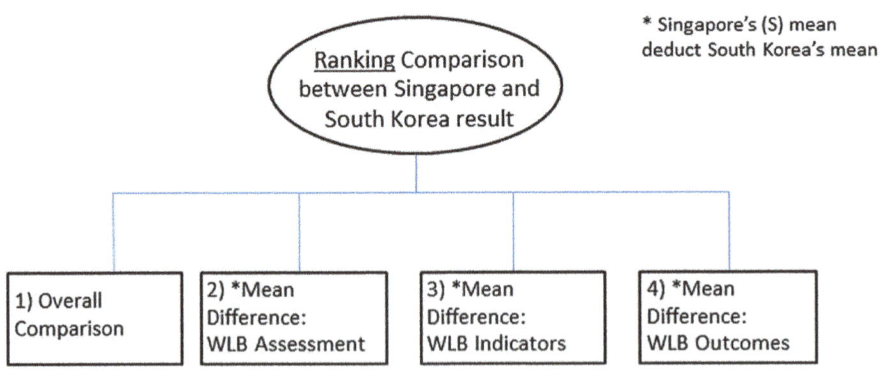

Fig. 5.2 Four-step comparison of the survey results between Singapore and South Korea

Step 4: WLB Outcomes—The WLB outcomes were items associated with the effects of attaining WLB. There were two indicators in this category: Work Outcomes and Life Outcomes. The mean difference in the scores between Singapore and South Korea for each item was also ranked and assessed correspondingly.

5.4.2 Inferential Statistics

It was critical to evaluate the conceptual framework with a software that is able to analyze the survey results comprehensively. As the conceptual framework consists of multiple variables that were measured simultaneously, statistical tools such as the SPSS do not have the interfaces to perform such an analysis. Hence, the Structural Equation Modeling (SEM) method was selected for this purpose. This method enabled the research team to apply sophisticated multivariate data analysis to produce a comprehensive report that interprets the relationships between multiple variables (Hair, Hult, Ringle, & Sarstedt, 2017). There are two types of SEM: Covariance-based SEM (CB-SEM) and Partial Least Squares SEM (PLS-SEM). For this current research, PLS-SEM (SmartPLS) was selected for several reasons. The PLS-SEM was primarily used to develop theories in exploratory research, it works efficiently with small sample sizes (<100), it has high efficiency in parameter estimation and the ability to support single item in a construct (Hair et al., 2017; Kim, 2016). All the above reasons were some of the more important criteria for choosing the PLS-SEM software. Hair, Ringle, and Sarstedt (2011: 139) defined PLS-SEM as "a causal modelling approach aimed at maximizing the explained variance of the dependent latent constructs." They also viewed the PLS-SEM as a silver bullet due to its robust statistical technology that could be applied to a much wider range of situations.

A series of assessment was conducted to evaluate the WLB conceptual framework and the proposed hypotheses. There are two parts to this assessment: assessment of the reflective measurement model and assessment of the structural model. Assessment of the reflective measurement model was conducted to establish the validity and reliability between the indicators and their respective sub-indicators. The tests used for assessing the measurement model include composite reliability, outer loadings, Average Variance Extracted (AVE), cross-loadings and the Fornell-Larcker criterion. After establishing the validity and reliability of the measurement model, an assessment of the structural model was conducted to examine the model's predictive capabilities and the relationships between the indicators (hypotheses). To assess the structural model, a six-step approach was executed: examining for collinearity issues, assessing the significance and relevancy of the structural model relationships, assessing the level of R^2, assessing the f^2 effect size, assessing the predictive relevance of Q^2 and assessing the q^2 effect size.

The questionnaire surveys conducted with the Singaporean and South Korean respondents were analyzed separately. The results of the SmartPLS model were used to compare with the research hypotheses formulated, the literature review and interview findings. Finally, a comparative assessment was also made between the results from Singapore and South Korea to identify any similarities or differences in the model which could be used for further study.

5.5 Summary

The survey and interview are two of the data collection methods used for analyzing the conceptual framework proposed in this study. Most survey questions were adapted from past research work while others were designed by the research team for this present study in order to suit the specific purpose of this research. Questionnaire surveys were conducted with the targeted respondents from Singapore and South Korea. The results of the survey were tested using statistical software that include SPSS and SmartPLS. Inferential statistics were used to test the hypotheses. The results of the survey analysis for Singapore and South Korea were used to craft the interview questions. The interviews were conducted to validate the survey findings as well as to gain qualitative insights of the results from the interviewees.

References

Balfour, D. L., & Wechsler, B. (1996). Organizational commitment: Antecedents and outcomes in public organizations. *Public Productivity & Management Review, 19*(3), 256–277.

Cammann, C., Fichman, M. K., & Williams, L. J. (1979). *The Michigan organizational assessment questionnaire*. Ann Arbor, Michigan: University of Michigan.

Carlson, D. S., Grzywacz, J. G., & Zivnuska, S. (2009). Is work–family balance more than conflict and enrichment? *Human Relations, 62*(10), 1459–1486. https://doi.org/10.1177/0018726709336500.

Carlson, D. S., Kacmar, K. M., Wayne, J. H., & Grzywacz, J. G. (2006). Measuring the positive side of the work–family interface: Development and validation of a work–family enrichment scale. *Journal of Vocational Behavior, 68*(1), 131–164. https://doi.org/10.1016/j.jvb.2005.02.002.

Carlson, D. S., Kacmar, K., & Williams, L. J. (2000). Construction and initial validation of a multidimensional measure of work-family conflict. *Journal of Vocational Behavior, 56*(2), 249–276. https://doi.org/10.1006/jvbe.1999.1713.

Diener, E., Emmons, R. A., Larsen, R. J., & Griffin, S. (1985). The satisfaction with life scale. *Journal of Personality Assessment, 49*(1), 71–75.

Gadd, G. (2012). Career enhancement through personal development. *Veterinary Nursing Journal, 27*(6), 235–236. https://doi.org/10.1111/j.2045-0648.2012.00183.x.

Haddon, B., & Hede, A. (2010). *WLB—An integrated approach*. University of the Sunshine Coast.

Hair, J. F., Hult, G. T. M., Ringle, C. M., & Sarstedt, M. (2017). *A primer on partial least squares structural equation modeling* (2nd ed.). Thousand Oaks: Sage.

Hair, J. F., Ringle, C. M., & Sarstedt, M. (2011). PLS-SEM: Indeed a silver bullet. *The Journal of Marketing Theory and Practice, 19*(2), 139–152. https://doi.org/10.2753/mtp1069-6679190202.

Hall, D. T., & Richter, J. (1988). Balancing work life and home life: What can organizations do to help? *Academy of Management Executive, 2*(3), 213–223. https://doi.org/10.5465/ame.1988.4277258.

Kim, G. S. (2016). Partial least squares structural equation modeling (PLS-SEM): An application in customer satisfaction research. *International Journal of U- and E-Service, Science and Technology, 9*(4), 61–68. http://dx.doi.org/10.14257/ijunesst.2016.9.4.07.

Lee, K. C. (2014). Factors that influence the decisions of working adults to enter the construction industry. Unpublished undergraduate dissertation, School of Design and Environment, National University of Singapore.

Lingard, H., & Francis, V. (2009). *Managing WLB in construction*. Hoboken: Taylor and Francis.

Lodahl, T. M., & Kejner, M. (1965). The definition and measurement of job involvement. *Journal of Applied Psychology, 49,* 24–33.

Mcmillan, H. S., Morris, M. L., & Atchley, E. K. (2010). Constructs of the work/life interface: A synthesis of the literature and introduction of the concept of work/life harmony. *Human Resource Development Review, 10*(1), 6–25. https://doi.org/10.1177/1534484310384958.

Quinn, R., & Staines, G. (1979). *The 1977 quality of employment survey: Descriptive statistics, with comparison data from the 1969-70 and the 1972-73 surveys*. Ann Arbor: University of Michigan, Survey Research Center.

Rizzo, J. R., House, R. J., & Lirtzman, S. I. (1970). Role conflict and ambiguity in complex organizations. *Administrative Science Quarterly, 15,* 150–168.

Ryff, C. D. (1989). Happiness is everything, or is it? Explorations on the meaning of psychological well-being. *Journal of Personality and Social Psychology, 57,* 1069–1081.

Seashore, S. E., Lawler, E. E., Mirvis, P. H., & Cammann, C. (1982). *Observing and measuring organizational change: A guide to field practice*. New York: Wiley.

Spreitzer, G. M. (1995). Psychological empowerment in the workplace: Dimensions, measurement, and validation. *Academy Management Journal, 38,* 1442–1465.

Valcour, P. M., & Batt, R. (2003). Work–life integration: Challenges and organizational responses. In P. Moen (Ed.), *It's about time: Couples and careers*. Ithaca, NY: Cornell University Press.

Voydanoff, P. (2005). Toward a conceptualization of perceived work-family fit and balance: A demands and resources approach. *Journal of Marriage and Family, 67*(4), 822–836. https://doi.org/10.1111/j.1741-3737.2005.00178.x.

Wayne, S. J., Shore, L. M., & Liden, R. C. (1997). Perceived organizational support and leader-member exchange: A social exchange perspective. *Academy of Management Journal, 40,* 82–111.

Zabriskie, R. B., & Ward, P. J. (2013). Satisfaction with family life scale. *Marriage & Family Review, 49*(5), 446–463. https://doi.org/10.1080/01494929.2013.768321.

Chapter 6
Research Findings and Analysis

Abstract This major chapter presents findings of the descriptive and inferential statistics. The descriptive statistics rank the Singapore and South Korea survey results to evaluate which attribute has the highest or lowest impact to the individual. The inferential statistics explain the Singapore and South Korea survey results anchored on the conceptual framework. Structural equation modelling performs the multivariate analysis on the conceptual framework to examine the relationships between variables. This chapter presents the validity and reliability of the measurement model as well as test results of the hypotheses. It also presents the interviewees' views and opinions of the findings from the statistical analysis.

6.1 Overview of Chapter

This chapter discusses the survey results using the appropriate software to perform the statistical analysis. Descriptive and inferential statistics were used to assimilate all the various WLB indicators and to draw meaningful explanations, using the software's interfaces, from the survey results. Verbatim reports from both the Singaporean and South Korean interviewees were also prepared to justify the statistical outcomes as well as to provide further insights if the results of the statistical analysis were different from observations of the literature review.

6.2 Survey Response and Demographics

As the demographics of the respondents were predetermined for this research, mass sending of survey questionnaires were not effective because there would be respondents who do not meet the criteria. For the survey in Singapore, emails were sent to main contractors (grade A1–B2 builders using the directory of the Building and Construction Authority) requesting them to invite their relevant employees to participate in the questionnaire survey. Out of 84 emails sent, eleven firms replied

L. Sui Pheng and B. K. Q. Chua, *Work–Life Balance in Construction*,
Management in the Built Environment,
https://doi.org/10.1007/978-981-13-1918-1_6

positively. Thirty-six completed survey questionnaires (including eight referrals from five respondents) were collected. However, three of the survey questionnaires were rejected because the respondents did not meet some criteria. For the survey in South Korea, due to communication problems, it was difficult for the research team to collect the data. Instead, the research team has requested one university based in Seoul for assistance. Survey questionnaires were given to the professors and students who have friends or family members belonging to the targeted group of respondents. Forty-four completed survey questionnaires were collected, all of which met the criteria stated. Based on the names of the companies written by all the respondents in the questionnaires, it could be noted that the pool of respondents came from seven different construction organizations in South Korea.

6.3 Descriptive Statistics

Table 6.1 shows the mean scores of all the survey questionnaires (before ranking). There were 111 questions in three categories: WLB assessment, WLB indicators and WLB outcomes. In this section, a comprehensive analysis of the mean scores from Singapore and South Korea was conducted to assess the similarities and/or differences between both countries. In addition, Table 6.1 also shows the survey questions together with their respective indicator names. However, in the subsequent analysis, only the indicator name would be shown. The reader can refer back to Table 6.1 or Appendix E to view the survey questions. Finally, the complete "indicator name" is shown in Table 6.2 for ease of understanding by the reader.

As the survey consisted of multi-item measurement scales, it was essential to incorporate reversed-polarity items to control and identify acquiescence response bias (Herche & Engelland, 1996). This could help reduce the all "yea-saying" or "nay-saying tendencies" (Herche & Engelland, 1996). Hence, in the survey questions, this study has preserved a mixture of positively coded and negatively coded items. The following indicators were negatively coded due to the negative impact these have to an individual: Work Demands (WD1–WD4), Boundary Spanning Demands, Life Demands (LD1–LD5), Work–Life Conflict and WO5. An example of a negatively coded item is: "I work under incompatible policies and guidelines". If this item has a score of 4 (Agree), it should be reverse-coded to 2 (Disagree) as though the question was asking "I work under compatible policies and guidelines". This means that an agreement to "I work under incompatible policies and guidelines" is logically similar to a disagreement to "I work under compatible policies and guidelines".

Before computing the mean scores for all items, reverse coding of the negative items must be conducted to ensure the mean scores reflected are all "positively" coded. This also allowed for a consistent assessment when ranking and comparisons were made between each item and countries. In this case, the SPSS software was used to reverse the coding of the negative items whereby 1 = 5, 2 = 4, 3 = 3, 4 = 2, and 5 = 1.

Table 6.1 Descriptive statistics of Singapore and South Korea before ranking

No.	Original order of the survey		Singapore		South Korea	
	Question	Indicator name	Mean	Std. Dev.[a]	Mean	Std. Dev.[a]
1	I do things that are apt to be accepted by one person and not accepted by others	WD1	**2.58**	1.091	**2.84**	1.055
2	I work under incompatible policies and guidelines	WD2	**2.97**	1.045	**3.11**	1.104
3	I do not know exactly what my responsibilities are at work	WD3	**3.61**	1.171	**4.18**	0.815
4	I am uncertain as to how my job is linked	WD4	**3.48**	0.972	**4.16**	1.010
5	I would like to spend more time in working	WD5	**2.24**	1.091	**3.77**	1.097
6	I have many roles in work that operate quite differently from one another	WD6	**3.33**	0.890	**2.68**	0.934
7	My job has often overload me with works that must be completed within a tight deadline	BSD1	**2.67**	1.190	**2.36**	1.080
8	Overloading with work in my job has negatively affected my personal life	BSD2	**2.61**	1.088	**2.48**	1.248
9	I often bring work to home as the project needs me to be available 24/7	BSD3	**3.45**	1.252	**3.39**	1.039
10	Bringing work to home has negatively affected my personal life	BSD4	**2.82**	1.261	**2.86**	1.374
11	My job allows me to bring life activities to work (e.g. Communicating with friends and family on phone)	BSD5	**3.18**	1.158	**2.77**	1.236
12	Bringing life activities to work has negatively affected my working life	BSD6	**3.48**	1.004	**3.59**	1.041
13	I often allow my personal life issues to be brought over to work	BSD7	**3.52**	1.004	**3.64**	1.036
14	I often get interrupted or distracted by my personal life issues at work	BSD8	**3.45**	1.003	**3.68**	1.006
15	I do things that are apt to be accepted by one person and not accepted by others	LD1	**3.15**	1.121	**3.09**	1.053
16	I have many roles in life that operate quite differently from one another	LD2	**2.73**	0.911	**2.70**	1.025
17	I do not know exactly what my responsibilities are in life	LD3	**3.52**	1.372	**4.14**	1.091
18	I do not know exactly what is expected of me in my life	LD4	**3.42**	1.347	**4.09**	1.053
19	I would not like to spend more time in my non-work life	LD5	**3.09**	1.588	**4.25**	1.059
20	The most important things that happen to me involve my life	LD6	**3.79**	0.927	**4.16**	0.963
21	The major satisfaction in my life comes from my job	WR1	**3.33**	1.137	**3.50**	1.110

(continued)

Table 6.1 (continued)

No.	Original order of the survey		Singapore		South Korea	
	Question	Indicator name	Mean	Std. Dev.[a]	Mean	Std. Dev.[a]
22	I am very much involved personally in my work	WR2	**3.27**	0.977	**3.25**	0.943
23	At the position that I have held at this organization, I have often been assigned projects that enabled me to develop and strengthen new skills	WR3	**3.33**	1.021	**3.20**	1.091
24	At the position that I have held at this organization, I have often been given additional challenging assignments	WR4	**3.30**	0.728	**3.43**	0.974
25	I have significant autonomy in determining how I do my job	WR5	**3.36**	0.783	**3.11**	0.945
26	I have considerable opportunity for independence and freedom in how I do my job	WR6	**3.58**	0.867	**2.95**	1.120
27	I am given flexible work schedule whereby I can work at a time that best suits me	BSR1	**2.82**	1.211	**2.41**	1.187
28	Flexible work schedule has positively affected my personal life	BSR2	**3.39**	1.144	**3.11**	1.166
29	My organization allowed me to apply time-off for non-work purposes	BSR3	**3.36**	1.319	**2.75**	1.164
30	Having time-off for non-work purposes has positively affected personal life	BSR4	**3.48**	1.121	**3.55**	1.044
31	I often participate in workshops for personal development	BSR5	**2.91**	1.156	**2.98**	1.151
32	Participating in workshops for personal development has positively affected my working and personal life	BSR6	**3.30**	0.847	**3.45**	0.975
33	I will assign time during my free days for work purposes	BSR7	**2.61**	1.144	**2.64**	1.143
34	Assigning time for work purposes have positively affected my working and personal life	BSR8	**2.97**	1.104	**2.59**	1.064
35	Most things in my life are more important than work	LR1	**3.88**	1.053	**3.50**	1.131
36	Most things in my life are more important than work	LR2	**3.73**	0.839	**3.95**	0.888
37	I feel that I am living in a harmonious life	LR3	**3.76**	0.708	**3.55**	1.066
38	My duty as a friend, family member or other non-work role is co-operating with each other harmoniously	LR4	**3.70**	0.883	**3.52**	0.876
39	I am satisfied with the recognition that I gotten from my families and friends	LR5	**3.70**	0.918	**3.73**	0.997

(continued)

Table 6.1 (continued)

No.	Original order of the survey		Singapore		South Korea	
	Question	Indicator name	Mean	Std. Dev.[a]	Mean	Std. Dev.[a]
40	I am satisfied with the compliments from my families and friends	LR6	**3.91**	0.947	**3.77**	0.859
41	I prefer taking up less demanding task than the work I am dealing right now	PMD1	**3.21**	1.111	**3.36**	0.967
42	I hope to shut off from work after 6 p.m.	PMD2	**3.97**	1.104	**4.30**	1.025
43	I hope my organization can play a bigger role in helping me cope with the high demand at work	PMD3	**3.73**	0.876	**4.11**	0.895
44	Reduce time with my love ones will allow me to better focus on my work	PMD4	**2.79**	1.341	**3.16**	1.180
45	Limit my usual lifestyle will allow me to have better focus on my work	PMD5	**3.00**	1.199	**2.57**	1.021
46	Reduce responsibility in my non-work life will help me attain balance between my work and life	PMD6	**3.03**	0.984	**2.64**	1.059
47	I want to have more challenging tasks assign to me	PMR1	**3.58**	1.062	**2.77**	1.097
48	I hope to have more breaks in between my working hour to allow me to relax my mind	PMR2	**3.55**	0.794	**3.84**	1.077
49	Having more freedom in my work allows me to produce better result	PMR3	**4.00**	0.935	**3.86**	1.069
50	I wish to take up new sports and hobbies to switch my mind off work	PMR4	**3.79**	0.960	**3.93**	1.065
51	Planning weekly activities with friends and family will allow me to have something to look forward to	PMR5	**3.94**	1.059	**4.16**	0.963
52	I want to have more uninterrupted personal time during my free time to keep away from work and life	PMR6	**3.67**	1.080	**3.89**	1.125
53	My work keeps me from my life activities more than I would like	WLCW1	**3.03**	1.015	**3.07**	1.108
54	The time I must devote to my job keeps me from participating equally in household responsibilities and activities	WLCW2	**3.15**	1.121	**2.89**	1.185
55	I have to miss life activities due to the amount of time I must spend on work responsibilities	WLCW3	**3.12**	0.960	**2.43**	1.021
56	When I get home from work I am often too frazzled to participate in life activities/ responsibilities	WLCW4	**2.67**	0.957	**2.95**	1.099
57	I am often so emotionally drained when I get home from work that it prevents me from contributing to my life	WLCW5	**2.85**	1.093	**2.91**	1.096

(continued)

Table 6.1 (continued)

No.	Original order of the survey		Singapore		South Korea	
	Question	Indicator name	Mean	Std. Dev.[a]	Mean	Std. Dev.[a]
58	Due to all the pressures at work, sometimes when I come home I am too stressed to do the things I enjoy	WLCW6	**2.79**	1.139	**2.95**	1.033
59	The problem-solving behaviors I use in my job are not effective in resolving problems at home	WLCW7	**3.15**	1.064	**2.77**	1.031
60	Behavior that is effective and necessary for me at work would be counterproductive at home	WLCW8	**2.85**	1.004	**3.23**	1.075
61	The behaviors I perform that make me effective at work do not help me to be a better person in my non-work life	WLCW9	**3.12**	1.139	**3.30**	1.047
62	The time I spend on life responsibilities often interfere with my work responsibilities	WLCL1	**3.30**	0.847	**3.25**	1.059
63	The time I spend with my life often causes me not to spend time in activities at work that could be helpful to my career	WLCL2	**3.09**	0.980	**3.23**	1.179
64	I have to miss work activities due to the amount of time I must spend on life responsibilities	WLCL3	**3.42**	1.062	**3.55**	1.150
65	Due to stress at home, I am often preoccupied with life matters at work	WLCL4	**3.24**	1.173	**3.50**	1.210
66	Because I am often stressed from life responsibilities, I have a hard time concentrating on my work	WLCL5	**3.73**	0.876	**3.48**	1.151
67	Tension and anxiety from my life often weakens my ability to do my job	WLCL6	**3.42**	1.001	**2.98**	1.210
68	The behaviors that work for me at home do not seem to be effective at work	WLCL7	**3.18**	1.014	**3.30**	1.047
69	Behavior that is effective and necessary for me at home would be counterproductive at work	WLCL8	**3.00**	0.935	**3.32**	1.137
70	The problem-solving behavior that works for me at home does not seem to be as useful at work	WLCL9	**3.42**	1.091	**3.50**	1.000
71	Helps me to understand different viewpoints and this helps me be a better person in my life	WLEW1	**3.79**	.927	**3.68**	0.883
72	Helps me to gain knowledge and this helps me be a better person in my life	WLEW2	**3.73**	1.008	**3.82**	0.843
73	Helps me acquire skills and this helps me be a better person in my life	WLEW3	**3.79**	1.023	**3.86**	0.878

(continued)

Table 6.1 (continued)

No.	Original order of the survey		Singapore		South Korea	
	Question	Indicator name	Mean	Std. Dev.[a]	Mean	Std. Dev.[a]
74	Puts me in a good mood and this helps me be a better person in my life	WLEW4	**3.33**	0.990	**3.52**	0.902
75	Makes me feel happy and this helps me be a better person in my life	WLEW5	**3.39**	0.899	**3.48**	1.000
76	Makes me cheerful and this helps me be a better person in my life	WLEW6	**3.48**	0.834	**3.57**	0.950
77	Helps me feel personally fulfilled and this helps me be a better person in my life	WLEW7	**3.61**	0.998	**3.86**	0.878
78	Provides me with a sense of accomplishment and this helps me be a better person in my life	WLEW8	**3.48**	0.939	**3.89**	0.813
79	Provides me with a sense of success and this helps me be a better person in my life	WLEW9	**3.64**	1.141	**3.84**	0.834
80	Helps me to gain knowledge and this helps me be a better worker	WLEL1	**3.67**	1.021	**3.82**	0.922
81	Helps me acquire skills and this helps me be a better worker	WLEL2	**3.67**	0.890	**3.80**	0.978
82	Helps me expand my knowledge of new things and this helps me be a better worker	WLEL3	**3.94**	0.609	**3.93**	0.998
83	Puts me in a good mood and this helps me be a better worker	WLEL4	**3.61**	1.059	**3.89**	0.895
84	Makes me feel happy and this helps me be a better worker	WLEL5	**3.52**	0.972	**3.86**	0.878
85	Makes me cheerful and this helps me be a better worker	WLEL6	**3.61**	1.029	**3.89**	0.813
86	Requires me to avoid wasting time at work and this helps me be a better worker	WLEL7	**3.61**	1.059	**3.64**	0.892
87	Encourages me to use my work time in a focused manner and this helps me be a better worker	WLEL8	**3.58**	0.936	**4.07**	0.661
88	Causes me to be more focused at work and this helps me be a better worker	WLEL9	**3.76**	1.001	**3.95**	0.746
89	Support from my colleagues and employer have helped me better cope with stress and improve my performance in work and life roles	CS1	**3.70**	0.951	**3.66**	0.987
90	Support from my friends and families have helped me better cope with stress and improve my performance in work and life roles	CS2	**3.76**	0.969	**4.05**	0.888
91	My usual leisure activities have helped me better cope with stress and improve my performance in work and life roles	CS3	**3.91**	0.723	**3.95**	0.914

(continued)

Table 6.1 (continued)

No.	Original order of the survey		Singapore		South Korea	
	Question	Indicator name	Mean	Std. Dev.[a]	Mean	Std. Dev.[a]
92	I am used to the usual work and life habits that make me have the ability to deal with high stress or low performance at work or life	CS4	**3.55**	0.794	**3.64**	0.942
93	I will engage in specific behavior which is appropriate to deal with high stress or low performance at work or life	CS5	**3.42**	1.146	**3.55**	0.875
94	I am able to negotiate and accomplish what is expected of me at work and in my life	WLB1	**3.58**	1.032	**3.77**	0.711
95	I am able to accomplish the expectations that my superior and my family have for me	WLB2	**3.48**	1.064	**3.84**	0.834
96	I do a good job of meeting the role expectations of critical people in my work and life	WLB3	**3.67**	0.854	**3.84**	0.680
97	My co-workers and friends and family would say that I am meeting their expectations	WLB4	**3.70**	0.810	**3.93**	0.789
98	People who are close to me would say that I do a good job of balancing work and life	WLB5	**3.52**	1.034	**3.57**	1.108
99	It is clear to me, based on feedback from co-workers, friends and family members, that I am accomplishing both my work and life responsibilities	WLB6	**3.58**	0.708	**3.70**	0.930
100	The work I do on my job is meaningful to me	WO1	**3.61**	0.899	**3.93**	0.950
101	All in all, I am satisfied with my job	WO2	**3.39**	0.933	**3.57**	0.998
102	I am proud to be able to tell people who it is I work for	WO3	**3.48**	0.834	**3.80**	0.978
103	I really feel as if my organization's problem is my own.	WO4	**3.27**	1.153	**3.32**	1.029
104	It is likely that I will be actively looking for new job next year	WO5	**2.79**	1.083	**2.98**	1.285
105	I will not give up on this job easily	WO6	**3.39**	1.116	**3.98**	.927
106	The conditions of my family life are excellent	LO1	**3.48**	1.064	**3.95**	1.120
107	I am satisfied with my family life	LO2	**3.76**	1.032	**4.02**	1.089
108	I am satisfied with my life	LO3	**3.55**	.833	**3.95**	1.077
109	In most ways my life is close to my ideal	LO4	**3.18**	1.044	**3.36**	1.080
110	In general, I feel confident and positive about myself	LO5	**3.52**	1.034	**3.93**	1.021
111	For the most part, I am proud of whom I am and the life I lead	LO6	**3.76**	0.969	**3.91**	0.910

Std. Dev.[a] = Standard Deviation

Table 6.2 Full name of each indicator

No.	Indicator name	Full name
1	WD	Work Demands
2	BSD	Boundary Spanning Demands
3	LD	Life Demands
4	WR	Work Resources
5	BSR	Boundary Spanning Resources
6	LR	Life Resources
7	PMD	Proposed Measures (Demand)
8	PMR	Proposed Measures (Resources)
9	WLCW	Work–Life Conflict (Work)
10	WLCL	Work–Life Conflict (Life)
11	WLEW	Work–Life Enrichment (Work)
12	WLEL	Work–Life Enrichment (Life)
13	CS	Coping Strategies
14	WLB	Work–Life Balance
15	WO	Work Outcomes
16	LO	Life Outcomes

6.4 Assessment of the Overall Ranked Mean Scores for Singapore

Table 6.3 shows the descriptive statistics of the overall WLB survey questionnaire for Singapore. Ranking of the mean scores for the entire WLB survey questionnaire was carried out to determine the relative importance of the various WLB variables as perceived by the respondents. The mean scores were ranked in descending order. In this section, the top 5 and bottom 5 mean scores were assessed.

6.4.1 Assessment of the Top 5 Overall Mean Scores for Singapore

Rank 1: PMR3: Having more freedom in my work allows me to produce better results

PMR3 belongs to the Proposed Measures indicator whereby implementing the proposed item would help to increase the WLE as well as decrease the WLC. PMR3 has the highest mean score of 4.00 that is equivalent to the respondents agreeing with this item. There were two reasons why Millennials Singaporeans desire freedom at work but might be constrained in doing so due to the prevailing culture in the society. Based on the Hofstede Cultural Dimensions, Singapore has a high score in Power Distance (74) which indicates that the power was not equally

Table 6.3 Ranked mean scores for all questions in the survey (Singapore)

Rank	Indicator name	Mean	Std. deviation	Rank	Indicator name	Mean	Std. deviation
1	PMR3	4.000	0.935	57	BSD3	3.455	1.252
2	PMD2	3.970	1.104	57	BSD8	3.455	1.003
3	PMR5	3.939	1.059	59	CS5	3.424	1.146
3	WLEL3	3.939	0.609	59	LD4	3.424	1.347
5	CS3	3.909	0.723	59	WLCL3	3.424	1.062
5	LR6	3.909	0.947	59	WLCL6	3.424	1.001
7	LR1	3.879	1.053	59	WLCL9	3.424	1.091
8	LD6	3.788	0.927	64	BSR2	3.394	1.144
8	PMR4	3.788	0.960	64	WLEW5	3.394	0.899
8	WLEW1	3.788	0.927	64	WO2	3.394	0.933
8	WLEW3	3.788	1.023	64	WO6	3.394	1.116
12	CS2	3.758	0.969	68	BSR3	3.364	1.319
12	LO2	3.758	1.032	68	WR5	3.364	0.783
12	LO6	3.758	0.969	70	WD6	3.333	0.890
12	LR3	3.758	0.708	70	WLEW4	3.333	0.990
12	WLEL9	3.758	1.001	70	WR1	3.333	1.137
17	LR2	3.727	0.839	70	WR3	3.333	1.021
17	PMD3	3.727	0.876	74	BSL6	3.303	0.847
17	WLCL5	3.727	0.876	74	WLCL1	3.303	0.847
17	WLEW2	3.727	1.008	74	WR4	3.303	0.728
21	CS1	3.697	0.951	77	WO4	3.273	1.153
21	LR4	3.697	0.883	77	WR2	3.273	0.977
21	LR5	3.697	0.918	79	WLCL4	3.242	1.173
21	WLB4	3.697	0.810	80	PMD1	3.212	1.111
25	PMR6	3.667	1.080	81	BSD5	3.182	1.158
25	WLB3	3.667	0.854	81	LO4	3.182	1.044
25	WLEL1	3.667	1.021	81	WLCL7	3.182	1.014
25	WLEL2	3.667	0.890	84	LD1	3.152	1.121
29	WLEW9	3.636	1.141	84	WLCW2	3.152	1.121
30	WD3	3.606	1.171	84	WLCW7	3.152	1.064
30	WLEL4	3.606	1.059	87	WLCW3	3.121	0.960
30	WLEL6	3.606	1.029	87	WLCW9	3.121	1.139
30	WLEL7	3.606	1.059	89	LD5	3.091	1.588
30	WLEW7	3.606	0.998	89	WLCL2	3.091	0.980
30	WO1	3.606	0.899	91	PMD6	3.030	0.984
36	PMR1	3.576	1.062	91	WLCW1	3.030	1.015
36	WLB1	3.576	1.032	93	PMD5	3.000	1.199
36	WLB6	3.576	0.708	93	WLCL8	3.000	0.935
36	WLEL8	3.576	0.936	95	BSL8	2.970	1.104

(continued)

Table 6.3 (continued)

Rank	Indicator name	Mean	Std. deviation	Rank	Indicator name	Mean	Std. deviation
36	WR6	3.576	0.867	95	WD2	2.970	1.045
41	CS4	3.545	0.794	97	BSR5	2.909	1.156
41	LO3	3.545	0.833	98	WLCW5	2.848	1.093
41	PMR2	3.545	0.794	98	WLCW8	2.848	1.004
44	BSD7	3.515	1.004	100	BSD4	2.818	1.261
44	LD3	3.515	1.372	100	BSR1	2.818	1.211
44	LO5	3.515	1.034	102	PMD4	2.788	1.341
44	WLB5	3.515	1.034	102	WLCW6	2.788	1.139
44	WLEL5	3.515	0.972	102	WO5	2.788	1.083
49	BSD6	3.485	1.004	105	LD2	2.727	0.911
49	BSR4	3.485	1.121	106	BSD1	2.667	1.190
49	LO1	3.485	1.064	106	WLCW4	2.667	0.957
49	WD4	3.485	0.972	108	BSD2	2.606	1.088
49	WLB2	3.485	1.064	108	BSR7	2.606	1.144
49	WLEW6	3.485	0.834	110	WD1	2.576	1.091
49	WLEW8	3.485	0.939	111	WD5	2.242	1.091
49	WO3	3.485	0.834				

distributed and this resulted in a strict hierarchy (Hofstede, 2001). In addition, the Singapore culture has always tended to be a Confucianism society and one characteristic of this is a hierarchically oriented society. With rigid organizations that do not exercise freedom for employees (control by management), this would result in employees desiring for freedom in their work.

Rank 2: PMD2: I hope to shut off from work after 6 p.m.

PMD2 also belongs to the Proposed Measures indicator whereby implementing the proposed item would help in increasing the WLE as well as decreasing the WLC. PMD2 has a mean score of 3.97 that suggests most of the respondents agreed with this statement. Employees in Singapore fall under the Employment Act as set out by legislation (Ministry of Manpower, 2011). Contractually, full time employees are not required to work more than 44 h in a week (Ministry of Manpower, 2011). However, due to the nature of the construction industry, it is very possible that employees are often held back at work to complete the tasks within the stipulated timeframe. Hence, this might be a reason why respondents hoped to shut themselves off from work after 6 p.m.

Rank 3: PMR5: Planning weekly activities with friends and family will allow me to have something to look forward to

Similar to the top two rankings, PMR5 belongs to the Proposed Measures indicator. PMR5 has a mean score of 3.94. It is no surprise for planning weekly activities with friends and families to have a high mean score. It is because this is one of the common ways where individuals can put aside their work to spend quality times with their loved ones. Many research studies found that individuals spend time with their friends or family members as a way to relieve their stress (Health Hub, 2015; Marshall, Allison, Nykamp, & Lanke, 2008; Opoku-Acheampong et al., 2017).

Rank 3: WLEL3: My involvement in life helps me expand my knowledge of new things and this helps me to be a better worker

WLEL3 belongs to the Work–Life Enrichment indicator where increasing the WLE can result in individual achieving Work–Life Balance. WLEL3 has a mean score of 3.94 that is similar to PMR5. Many studies have observed that work and life were unidirectional where work could interfere with life and vice versa (Carlson, Kacmar, & Williams, 2000). Carlson, Kacmar, Wayne, and Grzywacz (2006) have also noted that enrichment could occur when resources gained from one role directly or indirectly improved the performance of the other role. In Hofstede's Cultural Dimensions, Singapore has a high score (72) for Long Term Orientation (Hofstede, 2001). This indicates that Singaporeans were guided towards improving themselves as they were always preparing for the future. Seeking new ways (knowledge) to cope or adapt with their work culture clearly showed how Singaporeans were pragmatic toward their future.

Rank 5: LR6: I am satisfied with the compliments from my families and friends

LR6 belongs to the Life Resources indicator where increasing the resources available would result in improved WLE. This has a mean score of 3.91. It was found that psychological rewards were salient to individuals as these would boost their self-esteem, ego gratification and motivation (Voydanoff, 2005b). The high mean score for this item suggests that the respondents were satisfied with the compliments received from their loved ones. Hence, it appears that psychological rewards received from the individual's life role was sufficient and that it could help increase the performance in their other role domains (Voydanoff, 2005b).

Rank 5: CS3: My usual leisure activities have helped me better cope with stress and improve my performance in work and life roles

CS3 belongs to the Coping Strategies indicator where when individuals encountered high WLC or low WLE, they would find ways to cope and eventually achieve WLB. Similar to LR6, CS3 has a mean score of 3.91. A high score shows that Singaporean respondents were using leisure activities as an ideal way to cope with their stress or improve their performance with their work or life domain. Haddon, Hede, and Whiteoak (2009) also found there were a growing body of literature that proposed leisure activities as an effective way to maintain the health and wellness of employees. Moreover, leisure activities offered opportunities for rejuvenation, recovery and resilience in coping with stress (Haddon et al., 2009).

6.4.2 Assessment of the Bottom 5 Overall Mean Scores for Singapore

Rank 111: WD5: I would like to spend more time working

WD5 belongs to the Work Demands indicator where an increase in demand would lead to an increase in WLC and vice versa. WD5 has the lowest mean score of 2.24 which means many respondents disagreed with this statement. This item was not reverse-coded as the research team does not have a strong reason to justify that spending more time working was considered a negative item. The low mean score suggests that Singaporean respondents disagreed they would like to spend more time working. Research completed by Bannon, Ford and Meltzer (2011) found that Millennials employees do not want to repeat what their previous generation have done which they perceived to be a mistake to work long hours at the expense of family and friends. They believed that quality of work was more important than the number of hours spent at work (Bannon et al., 2011).

Rank 110: WD1: I do things that apt to be accepted by one person and not accepted by others

WD1 belongs to the Work Demands indicator where an increase in demand would lead to an increase in WLC and vice versa. WD1 has the second lowest mean score of 2.58. This suggests that most respondents disagreed with this statement. As WD1 has been reverse-coded to a positive item, this suggests that Singaporean respondents disagreed that they do not do things that were apt to be accepted by one but not accepted by another. This also means that they tended to agree that they do things that were accepted by one but not accepted by others. One possible reason for this phenomenon could be due to the nature of the construction industry where there were many players in a project and different stakeholders might have different opinions on the same issue. Therefore, it was very likely that work might be accepted by one but not accepted by others due to conflicts of opinions.

Rank 108: BSR7: I will assign time during my free days for work purposes

BSR7 belongs to the Boundary Spanning Resources indicator where an increase in BSR would lead to an increase in WLE. Even though BSR7 originated from one domain, it served as resources in both work and life domains (Voydanoff, 2005b). BSR7 has the third lowest mean score of 2.61 that means that on average, respondents disagreed with this statement. This was not the least surprising as this suggests that the Singaporean respondents cherished their free time to spend on their life purposes to rejuvenate from their long hours at work. In Hofstede's Cultural Dimension of Indulgence versus Restraint, Singapore has a close to mid-point score of 46 which signified that while enjoying the indulgence part of their life, Singaporeans also restrained themselves to abide with their strict social norms. This was where the "work hard, play hard" lifestyle was important for them to cope with high stress at work.

Rank 108: BSD2: Overloading with work in my job has negatively affected my personal life

BSD2 belongs to the Boundary Spanning Demands indicator where an increase in demand would lead to an increase in WLC and vice versa. Similar to BSR7, this originated from one domain but served as demands in both work and life domains (Voydanoff, 2005b). BSD2 has a mean score of 2.61 and it has been reverse-coded to a positive item. This means the Singaporean respondents disagreed that overloading them with work in their jobs could positively affect their personal life. This trend was not unexpected judging by the nature of the construction industry, where the amount of workload an individual has could be substantial especially when faced with tight deadlines (Lingard & Francis, 2009). This could lead to excess working hours needed to complete the work before the deadlines causing them to have less time to spend in their life roles. If this issue persisted, this might eventually lead to employee's intention to turnover (Lingard & Francis, 2009).

Rank 106: WLCW4: When I get home from work I am often too frazzled to participate in life activities/responsibilities

WLCW4 belongs to the Work–Life Conflict indicator where a decrease in WLC could result in an individual achieving Work–Life Balance. WLCW4 has a mean score of 2.67 and it was reverse-coded to a positive item. This implies that the Singaporean respondents disagreed that they felt energized to participate in their life domain after work. It was very common for respondents to disagree with this statement as work-hours were usually 8–10 h long excluding overtime. This could completely exhaust an individual especially during hectic days where there were much works to do. As a result, this issue generated strain that undermined an individual's ability to fulfill the responsibilities of their life roles (Greenhaus & Beutell, 1985).

Rank 106: BSD1: My job has overload me with works that must be completed within a tight deadline

BSD1 belongs to the Boundary Spanning Demands indicator where an increase in demand would lead to an increase in WLC and vice versa. BSD1 has a mean score of 2.67 and it was reverse-coded to a positive item. This suggests that on average, the Singaporean respondents disagreed that their jobs do not overload them with works that must be completed within tight deadlines. As mentioned in BSD2, due to the nature of the construction industry, it was unavoidable to have works that must be completed within tight deadlines causing stress to build up at work as well as their life after work. Given the low mean scores for BSD1 and BSD2, these suggest that Singaporean respondents were dissatisfied with the workload given to them and this has negatively affected their work and life domain.

6.5 Assessment of the Overall Ranked Mean Scores for South Korea

Table 6.4 shows the descriptive statistics of the overall WLB survey questionnaire for South Korea. Ranking of the mean scores for the entire WLB survey questionnaire was carried out to determine the relative importance of the various WLB variables as perceived by the respondents. The mean scores were ranked in descending order. In this section, the top 5 and bottom 5 mean scores were assessed and discussed.

Table 6.4 Ranked mean scores for all questions in the survey (South Korea)

Rank	Indicator name	Mean	Std. deviation	Rank	Indicator name	Mean	Std. deviation
1	PMD2	4.295	1.025	55	LR3	3.545	1.066
2	LD5	4.250	1.059	55	WLCL3	3.545	1.150
3	WD3	4.182	0.815	59	LR4	3.523	0.876
4	LD6	4.159	0.963	59	WLEW4	3.523	0.902
4	PMR5	4.159	0.963	61	LR1	3.500	1.131
4	WD4	4.159	1.010	61	WLCL4	3.500	1.210
7	LD3	4.136	1.091	61	WLCL9	3.500	1.000
8	PMD3	4.114	0.895	61	WR1	3.500	1.110
9	LD4	4.091	1.053	65	WLCL5	3.477	1.151
10	WLEL8	4.068	0.661	65	WLEW5	3.477	1.000
11	CS2	4.045	0.888	67	BSR6	3.455	0.975
12	LO2	4.023	1.089	68	WR4	3.432	0.974
13	WO6	3.977	0.927	69	BSD3	3.386	1.039
14	CS3	3.955	0.914	70	LO4	3.364	1.080
14	LO1	3.955	1.120	70	PMD1	3.364	0.967
14	LO3	3.955	1.077	72	WD6	3.318	0.934
14	LR2	3.955	0.888	72	WLCL8	3.318	1.137
14	WLEL9	3.955	0.746	72	WO4	3.318	1.029
19	LO5	3.932	1.021	75	WLCL7	3.295	1.047
19	PMR4	3.932	1.065	75	WLCW9	3.295	1.047
19	WLB4	3.932	0.789	77	WLCL1	3.250	1.059
19	WLEL3	3.932	0.998	77	WR2	3.250	0.943
19	WO1	3.932	0.950	79	WLCL2	3.227	1.179
24	LO6	3.909	0.910	79	WLCW8	3.227	1.075
25	PMR6	3.886	1.125	81	WR3	3.205	1.091
25	WLEL4	3.886	0.895	82	PMD4	3.159	1.180
25	WLEL6	3.886	0.813	83	BSR2	3.114	1.166
25	WLEW8	3.886	0.813	83	WD2	3.114	1.104

(continued)

Table 6.4 (continued)

Rank	Indicator name	Mean	Std. deviation	Rank	Indicator name	Mean	Std. deviation
29	PMR3	3.864	1.069	83	WR5	3.114	0.945
29	WLEL5	3.864	0.878	86	LD1	3.091	1.053
29	WLEW3	3.864	0.878	87	WLCW1	3.068	1.108
29	WLEW7	3.864	0.878	88	BSR5	2.977	1.151
33	PMR2	3.841	1.077	88	WLCL6	2.977	1.210
33	WLB2	3.841	0.834	88	WO5	2.977	1.285
33	WLB3	3.841	0.680	91	WLCW4	2.955	1.099
33	WLEW9	3.841	0.834	91	WLCW6	2.955	1.033
37	WLEL1	3.818	0.922	91	WR6	2.955	1.120
37	WLEW2	3.818	0.843	94	WLCW5	2.909	1.096
39	WLEL2	3.795	0.978	95	WLCW2	2.886	1.185
39	WO3	3.795	0.978	96	BSD4	2.864	1.374
41	LR6	3.773	0.859	97	WD1	2.841	1.055
41	WLB1	3.773	0.711	98	BSD5	2.773	1.236
43	LR5	3.727	0.997	98	PMR1	2.773	1.097
44	WLB6	3.705	0.930	98	WLCW7	2.773	1.031
45	BSD8	3.682	1.006	101	BSR3	2.750	1.164
45	WLEW1	3.682	0.883	102	LD2	2.705	1.025
47	CS1	3.659	0.987	103	BSL7	2.636	1.143
48	BSD7	3.636	1.036	103	PMD6	2.636	1.059
48	CS4	3.636	0.942	105	BSR8	2.591	1.064
48	WLEL7	3.636	0.892	106	PMD5	2.568	1.021
51	BSD6	3.591	1.041	107	BSD2	2.477	1.248
52	WLB5	3.568	1.108	108	WLCW3	2.432	1.021
52	WLEW6	3.568	0.950	109	BSR1	2.409	1.187
52	WO2	3.568	0.998	110	BSD1	2.364	1.080
55	BSR4	3.545	1.044	111	WD5	2.227	1.097
55	CS5	3.545	0.875				

6.5.1 Assessment of the Top 5 Overall Mean Scores for South Korea

Rank 1: PMD2: I hope to shut off after 6 p.m.

PMD2 belongs to the Proposed Measures indicator whereby implementing the proposed item would help individuals increase their WLE as well as decrease the WLC. In South Korea's assessment, PMD2 was ranked the highest with a mean score of 4.30 where 26 respondents out of 44 respondents have strongly agreed with this statement. This item was ranked second in Singapore's assessment. This clearly shows that shutting off from work at 6 p.m. was one of the most important items to achieving WLB. In Hofstede's Cultural Dimensions for Power Distance, South Korea has a score of 60, indicating that they followed the hierarchical structure in the Korean society and accepted power inequality among individuals. Bae (2017) has also mentioned that many Korean workers would still remain in the office after work hours as they have to wait for their superiors to leave. This suggests that South Koreans were still preserving the Confucian values of respecting seniority. However, the respondents still found it important to shut off work after 6 p.m. to spend more time in their life domain.

Rank 2: LD5: I would not like to spend more time in my non-work life

LD5 belongs to the Life Demand indicator where an increase in demand would increase the WLC of an individual and vice versa. LD5 was ranked second with a mean score of 4.25 and it was reverse-coded to a positive item. This suggests that on average, respondents agreed or strongly agreed that they would like to spend more time in their life domain. Millennials employees cherished their life activities and their free time was as important as their work time (Bannon et al., 2011). With South Koreans desiring to shut off work after 6 p.m. and hoping to spend more time in their life domains, this highlights the importance of WLB to the Millennials respondents.

Rank 3: WD3: I do not know exactly what my responsibilities are at work

WD3 belongs to the Work Demands indicator and has a mean score of 4.18. WD3 has been reverse-coded to a positive item that means on average, respondents agreed that they knew exactly what their responsibilities were. Knowing what their responsibilities were was important for individuals working in the construction industry. The durations of building projects were relatively short and project teams must therefore be assembled quickly and could not often be restructured in between major milestones (Low, 1992). Hence, it was important for project members to know what their responsibilities were prior to the start of the construction project.

Rank 4: LD6: The most important things that happen to me involve my life

LD6 belongs to the Life Demands indicator and has a mean score of 4.16. The high mean score of 4.16 means that on average, South Korean respondents agreed that the most important thing that happened to them involved their life. Aruna and Anitha (2015) opined that Millennials employees gave priorities to their friends and family members rather than work. Similar to LD5, LD6 also mentioned the importance of life involvement of an individual. These two items with high mean scores showed that the South Koreans placed high emphasis on life involvement rather than work involvement.

Rank 4: PMR5: Planning weekly activities with friends and family will allow me to have something to look forward to

PMR5 belongs to the Proposed Measures indicator and has a mean score of 4.16. This item was ranked third in Singapore's overall assessment that has a mean score of 3.94. The high mean scores clearly showed that PMR5 was an important measure for both Singaporeans and South Koreans because on average, most individuals agreed that planning weekly activities with their loved ones allowed them to have something to look forward to every week. Furthermore, it also helped them to reduce their WLC and improve their WLE. This trend also concurred with past research studies that found individuals spending time with their friends and family members as a way to relief stress (Health Hub, 2015; Marshall et al., 2008; Opoku-Acheampong et al., 2017).

Rank 4: WD4: I am uncertain as to how my job is linked

WD4 belongs to the Work Demands indicator and has a mean score of 4.16. WD4 was reverse-coded to a positive item which suggested that on average, the South Koreans agreed that they were certain as to how their jobs were linked. Another noteworthy observation was that WD3 and WD4 belonged to the same sub-indicator called Work Role Ambiguity. Work Role Ambiguity was used to measure the level of ambiguity an individual faced relating to their roles and responsibilities when performing a particular role (Michel, Kotrba, Mitchelson, Clark, & Baltes, 2010). The high mean score for this item as well as WD3 showed that the South Korean respondents were clear on how their jobs were linked and what their responsibilities were. As mentioned earlier, due to the nature of the construction industry, it was crucial for all project team members to understand what their role and responsibilities were. Moreover, in Hofstede's Cultural Dimensions, South Korea has a high score of 85 in Uncertainty Avoidance. This showed that South Koreans do not tolerate uncertainty and ambiguity in life and would follow what have been told to them to minimize the occurrence of the unknown (Hofstede, 2001). This could also be the reason why the South Koreans have a high mean score for this item.

6.5.2 Assessment of the Bottom 5 Overall Mean Scores for South Korea

Rank 111: WD5: I would like to spend more time working

WD5 belongs to the Work Demands indicator where an increase in demand would lead to an increase in WLC and vice versa. WD5 has the lowest mean score of 2.23 where 13 respondents strongly disagreed and 15 respondents disagreed with this statement. This trend was similar to Singapore's assessment where WD5 was also ranked last. This means that Singapore and South Korea Millennials employees would not want their work domain to exceed their life domain boundary. Once their work hours were over, they would want to completely shut themselves off from work. Moreover, with items like "I will like to spend more time in my non-work life" and "The most important things that happen to me involve my life" being ranked 2nd and 4th respectively in South Korea's assessment, the analysis showed that the South Koreans weighed their life domain more highly than their work domain.

Rank 110: BSD1: My job has often overload me with works that must be completed within tight deadline

BSD1 belongs to the Boundary Spanning Demands indicator and has a mean score of 2.36. This item has been reverse-coded to a positive item such that on average, South Korean respondents disagreed that their jobs do not overload them with works that must be completed within tight deadlines. As mentioned in the assessment for Singapore, due to the nature of the construction industry, it was unavoidable to have works that must be completed within tight deadlines. Lee (2014) found that employees overloaded with work was one of the main factors that contributed to the stress buildup and other physical health issues. Both Singapore and South Korea ranked this item lowly at 106 and 110 respectively. Hence, this issue must be examined by the relevant authorities or organizations to help the employees reduce any negative impact that is associated with work overload.

Rank 109: BSR1: I am given flexible work schedule whereby I can work at a time that best suits me

BSR1 belongs to the Boundary Spanning Resources indicator and has a mean score of 2.41. The low mean score for BSR1 suggests that the South Korean respondents disagreed that they were given flexible work schedules that best suit them. Lee (2014) mentioned that working hours in the construction industry have always been perceived as long and inflexible. Furthermore, as construction projects involved many different project teams or stakeholders, it might be difficult for one to accommodate other's flexible work schedules especially during site meetings where every project team members must be involved. Hence, although flexible work schedule was highly desired, it might be challenging to implement this in the construction industry.

Rank 108: WLCW3: I have to miss life activities due to the amount of time I must spend on work responsibilities

WLCW3 belongs to the Work–Life Conflict indicator and has a mean score of 2.43. WLCW3 has been reverse-coded to a positive item, meaning that on average, respondents disagreed that they need not have to miss their life activities due to the amount of time they spent at work. As mentioned earlier, working hours in the construction industry were long and inflexible (Lee, 2014). Furthermore, the above analysis showed that on average, the South Korean respondents were overloaded with work that must be completed within tight deadlines. Hence, it could be deduced that these respondents have to miss their life activities due to the amount of time spent at work.

Rank 107: BSD2: Overloading with work in my job has negatively affected my personal life

BSD2 belongs to the Boundary Spanning Demands indicator and has a mean score of 2.48. BSD2 has been reverse-coded to a positive item. This means the South Korean respondents disagreed that overloading with work in their jobs has positively affected their personal life. As seen in South Korea's BSD1, WD5, and WLCW3 explanations, it was obvious that the respondents were overloaded with work that have negatively affected their life domain judging by the low mean scores. This issue not only happened in South Korea. The assessment of Singapore has also showed that both BSD1 and BSD2 ranked within the bottom five items. This suggests that both countries faced the same issue despite their cultural differences and construction practices. Even though the nature of the construction industry has made overloading at work an unavoidable issue, finding ways to reduce this problem would surely help individuals reduce stress buildup and other physical health problems. Hence, as mentioned earlier, it is critical for the relevant authorities and organizations to look for ways to mitigate this issue.

6.6 Assessment of the Mean Differences Between Singapore and South Korea

In this section, the mean differences between Singapore and South Korea's scores were computed and compared. There are three different assessments in this part: WLB assessment, WLB indicators and WLB outcomes. WLB assessments are items that precede and contribute to the overall appraisal of WLB. WLB indicators are items used to represent the individual's level of WLB. The WLB outcomes are items that succeed the WLB indicator and used to assess the WLB outcomes of an individual. Items from these three assessments were ranked separately and the highest or lowest mean difference from each assessment were analyzed. It is important to note that the scores were ranked in descending order and the formula used for computing the mean difference is:

$$Singapore\,Mean\,Score - South\,Korea\,Mean\,Score = Mean\,Difference$$

6.7 Assessment of the Work–Life Balance Assessment Between Singapore and South Korea

In this analysis, the mean differences for WLB assessment consisting 93 items were ranked and the highest (top 5) and lowest (bottom 5) scores compared and assessed. The ten indicators that served as the WLB assessment based on the conceptual framework are as follows: Work Demands, Boundary Spanning Demands, Life

Demands, Work Resources, Boundary Spanning Resources, Life Resources, Proposed Measures, WLE, WLC and Coping Strategies. Table 6.5 presents the mean difference rankings of the complete WLB assessment in descending order.

6.7.1 Assessment of the Top 5 Items in Work–Life Balance Assessment

Rank 1: PMR1: I want to have more challenging tasks assigned to me

PMR1 belongs to the Proposed Measures indicator and assigning challenging task to an individual was seen as a strategy to help individuals increase their resources that eventually lead to increase in WLE. As mentioned earlier, individuals pursuing promotion in the organization would be more interested in receiving more challenging tasks to support their career advancement (Lingard & Francis, 2009). Based on Tables 6.6 and 6.7, Singapore has a mean score of 3.58 and South Korea has a mean score of 2.77. For Singapore, more than 60% of the respondents agreed that they want to have more challenging tasks assigned to them as compared to South Korea where only 25% of the respondents agreed with this statement. This could be related to Hofstede's cultural dimension of Uncertainty Avoidance. Singapore has a low score of 8 for this dimension meaning that the Singaporeans accepted uncertainty and ambiguity in life and tended to be more relaxed when there were opportunities given to them (Hofstede, Hofstede, & Minkov, 2010). In contrast, South Korea has a high score of 85 for this dimension indicating that the South Koreans tended to avoid uncertainty and ambiguity and valued security in their jobs. This dimension reflects the situation where Singaporeans were more willing to accept challenging tasks assigned to them even though there might be many uncertainties as to whether they would complete the tasks on time etc. On the other hand, South Koreans might be more conservative as they might worry that there were too many uncertainties and would try to avoid the tasks by all necessary means.

Rank 2: WLCW3: I have to miss my life activities due to the amount of time I spend on work responsibilities

WLCW3 belongs to the Work–Life Conflict indicator. It is a form of inter-role conflict in which participation in work and life domains are seen as mutually incompatible. WLCW3 was a time-based conflict where time allocation in work responsibilities hindered the role performance in the life domain (Michel et al., 2010). It is important to note that WLCW3 was reverse-coded to a positive item, meaning the results for WLCW3 shown in Tables 6.6 and 6.7 were reversed. WLCW3 has a mean difference of 0.69 and the mean scores for Singapore and South Korea were 3.12 and 2.43 respectively. Based on the breakdown shown, it could be seen that close to 66% of the South Korean respondents disagreed that they have not miss their life activities due to the amount of time spent at work.

Table 6.5 Mean difference of Work–Life Balance assessment between Singapore and South Korea

Ranking	Indicators	Mean difference	Ranking	Indicators	Mean difference
1	PMR1	0.803	48	WLEW2	−0.091
2	WLCW3	0.689	49	BSD6	−0.106
3	WR6	0.621	50	WLCL7	−0.114
4	BSR3	0.614	51	BSD7	−0.121
5	WLCL6	0.447	51	WLCL3	−0.121
6	PMD5	0.432	51	CS5	−0.121
7	BSR1	0.409	54	WR4	−0.129
8	BSD5	0.409	55	WLEL2	−0.129
9	PMD6	0.394	56	WLCL2	−0.136
10	BSR8	0.379	57	PMR4	−0.144
10	LR1	0.379	58	WD2	−0.144
12	WLCW7	0.379	59	BSR6	−0.152
13	BSD1	0.303	60	PMD1	−0.152
14	BSR2	0.280	60	WLEL1	−0.152
15	WLCW2	0.265	62	WR1	−0.167
16	WR5	0.250	62	WLCW6	−0.167
16	WLCL5	0.250	64	WLCW9	−0.174
18	LR3	0.212	65	WLEW4	−0.189
19	LR4	0.174	66	WLEL9	−0.197
20	LR6	0.136	67	WLEW9	−0.205
20	PMR3	0.136	68	PMR5	−0.220
22	BSD2	0.129	68	PMR6	−0.220
22	WR3	0.129	70	BSD8	−0.227
24	WLEW1	0.106	71	LR2	−0.227
25	BSD3	0.068	72	WLCL4	−0.258
26	LD1	0.061	72	WLEW7	−0.258
27	WLCL1	0.053	74	WD1	−0.265
28	CS1	0.038	75	WLEL4	−0.280
29	WR2	0.023	75	WLEL6	−0.280
30	LD2	0.023	77	WLCW4	−0.288
31	WD5	0.015	77	CS2	−0.288
31	WD6	0.015	79	PMR2	−0.295
33	WLEL3	0.008	80	WLCL8	−0.318
34	BSR7	−0.030	81	PMD2	−0.326
34	LR5	−0.030	82	WLEL5	−0.348
34	WLEL7	−0.030	83	LD6	−0.371
37	WLCW1	−0.038	83	PMD4	−0.371
38	BSD4	−0.045	85	WLCW8	−0.379
38	CS3	−0.045	86	PMD3	−0.386

(continued)

Table 6.5 (continued)

Ranking	Indicators	Mean difference	Ranking	Indicators	Mean difference
40	BSR4	−0.061	87	WLEW8	−0.402
40	WLCW5	−0.061	88	WLEL8	−0.492
42	BSR5	−0.068	89	WD3	−0.576
43	WLCL9	−0.076	90	LD3	−0.621
43	WLEW3	−0.076	91	LD4	−0.667
45	WLEW5	−0.083	92	WD4	−0.674
46	WLEW6	−0.083	93	LD5	−1.159
47	CS4	−0.091			

Table 6.6 A breakdown of the mean scores for the top 5 items (Singapore)

Singapore	Mean score	Strongly disagree	Disagree	Neither agree nor disagree	Agree	Strongly agree
PMR1	3.58	1	5	7	14	6
%		3.0%	15.2%	21.2%	42.4%	18.2%
WLCW3	3.12	0	11	9	11	2
%		0.0%	33.3%	27.3%	33.3%	6.1%
WR6	3.58	1	3	7	20	2
%		3.0%	9.1%	21.2%	60.6%	6.1%
BSR3	3.36	5	3	6	13	6
%		15.2%	9.1%	18.2%	39.4%	18.2%
WLCL6	3.42	1	6	7	16	3
%		3.0%	18.2%	21.2%	48.5%	9.1%

Table 6.7 A breakdown of the mean scores for the top 5 items (South Korea)

South Korea	Mean score	Strongly disagree	Disagree	Neither agree nor disagree	Agree	Strongly agree
PMR1	2.77	7	9	17	9	2
%		15.9%	20.5%	38.6%	20.5%	4.5%
WLCW3	2.43	5	24	9	3	3
%		11.4%	54.5%	20.5%	6.8%	6.8%
WR6	2.95	4	14	8	16	2
%		9.1%	31.8%	18.2%	36.4%	4.5%
BSR3	2.75	7	12	13	9	3
%		15.9%	27.3%	29.5%	20.5%	6.8%
WLCL6	2.98	4	13	14	6	7
%		9.1%	29.5%	31.8%	13.6%	15.9%

Singapore has balanced scores where 33% of the respondents disagreed and agreed and 27.3% neither agreed nor disagreed. The target therefore naturally shifted towards South Korea because the proportion of disagreement was an alarming figure. Based on a study by the Organization for Economic Co-operation and Development, the South Korean worker was ranked among those with one of the highest annual working hours (2,113 h) as compared to the average OECD countries (1,766 h) (Organization for Economic Co-operation and Development, 2016). This could also be explained by Hofstede's cultural dimension of Power Distance where South Korea has a score of 60. This suggests that South Koreans followed a strict hierarchical structure in their society especially in the workplace where employees would not leave the office until their superiors have left the workplace (Bae, 2017; Olsen, 2008). Furthermore, with South Korea having a high score in Hofstede's cultural dimension of Uncertainty Avoidance, the South Koreans would tend to avoid uncertainties. This means they would use formal rules to regulate their behavior and to follow what have been viewed as norms in their society. As a result, Millennials employees would also continue to follow the norms that have been passed down for generations. This was unlike Singapore that has a low Uncertainty Avoidance score, even though the Power Distance score was high. Hence, in Singapore's culture, Singaporeans would tend to accept uncertainties and would look for opportunities to change their current work environment to fit into their life domain.

Rank 3: WR6: At the position that I have held at this organization, I have often been assigned projects that enabled me to develop and strengthen new skills

WR6 belongs to the Work Resources indicator that would increase resources to individuals when the organization provides them with developmental opportunities. WR6 has a mean difference score of 0.62 (after rounding off) and the mean scores for Singapore and South Korea were 3.58 and 2.95 respectively. Based on Tables 6.6 and 6.7, it could be seen that close to 67% of the Singaporean respondents agreed that the organization has often assigned projects that enabled them to develop and strengthen new skills. However, only 13% of the South Korean respondents agreed with this item. This reflected the opportunities given by the construction firms in the two countries. Based on Hofstede's cultural dimensions, although Singapore belonged to a power inequality society where the Singaporeans followed a strict hierarchical structure, they also belonged to a collectivized society. This suggests that Singaporeans preferred working as a team and they trusted their team members to deliver quality results. Furthermore, Singaporeans accepted uncertainties and ambiguity given to them and they tended to be more relaxed and were open to changes given to them (Hofstede, 2001). Collectively, this shows that top management was more open to changes and was more willing to give their subordinates a chance to develop their new skills. Even though South Korea was a society focused on collectivism, South Koreans were more unlikely to accept uncertainties and ambiguity in life. Hence, this suggests that superiors in South

Korea were less likely to give subordinates a chance to try out new assignments that could help them develop their skills because the level of uncertainties was high.

Rank 4: BSR3: My organization allowed me to apply time-off for non-work purposes

BSR3 belongs to the Boundary Spanning Resources indicator where organizations allowed individuals to apply time-off for their life purposes. This would help individuals increase their resources in their work and life domain. BSR3 has a mean difference of 0.61 and the mean scores for Singapore and South Korea for this item were 3.36 and 2.75 respectively. Based on Tables 6.6 and 6.7, it could be seen that more than 57% of the Singaporean respondents agreed that their organizations allowed them to apply time off for life purposes as compared to South Korea that only has 27%. This could be explained by Hofstede's cultural dimension of Indulgence versus Restraint. South Korea has a low score of 29 that suggested the South Koreans were restrained by strict social norms and have a perception that indulgence was somewhat wrong. It might seem that the superiors were more unlikely to allow their subordinates to apply time-off for non-work purposes as they perceived it as a way to avoid work. In addition, Olsen (2008) mentioned that South Korea's culture of hard work was deeply ingrained into the society and employees presumed that leaving work on time or early might not get them a promotion or pay rise. This could also be the reason why they felt their organizations do not allow them to apply time-off for non-work purposes. On the other hand, Singapore has a low score in the Uncertainty Avoidance dimension that suggested the Singaporean culture tended to be in a relaxed attitude and willing to open to changes. In a recent study by Singapore's Ministry of Manpower, it was found that organizations were practicing beyond the statutory requirements in leave benefits to help employees cope with personal and family commitments (Channel NewsAsia, 2016). Furthermore, the study showed that 75% of the firms surveyed gave their employees unplanned time-off for non-work purposes. This reflected Singaporean organizations' willingness to help their employees cope with life commitments as well as stress developed at work.

Rank 5: WLCL6: Tension and anxiety from my life often weaken my ability to do my job

WLCL6 belongs to the Work–Life Conflict indicator and was viewed as a strain-based conflict. An increase in tension or anxiety in the life domain could hinder the role performance in the work domain (Michel et al., 2010). It is important to note that WLCL6 was reverse-coded to a positive item ("Tension and anxiety from my life does not weaken my ability to do my job"), meaning the results for WLCL6 shown in Tables 6.6 and 6.7 were reversed from 5 to 1, 4 to 2, 3 to 3, 2 to 4 and 1 to 5. The mean difference for WLCL6 was 0.45 (with rounding off) and the mean scores for Singapore and South Korea were 3.42 and 2.98 respectively. Based on the breakdown shown in Tables 6.6 and 6.7, close to 58% of the Singaporean respondents agreed that tension and anxiety from their life does not often weaken their ability to do their jobs while only 20% disagreed to this statement. This suggests that Singaporeans were more segmental where work and life domains were deemed to be mutually exclusive with their distinct mentality. This also means individual's life related matters were not brought into the work domain as they might not want to affect their job performance. On the other hand, the results from South Korea were not obvious because 38% of the respondents disagreed with this statement versus 29% who agreed with this statement. Hence, this study was unable to justify which end of the continuum (ranging from segmentation to integration) the South Koreans belong. However, based on Hofstede's cultural dimension on Indulgence versus Restraint, it could be deduced that South Koreans have the tendency to look at life negatively and always believed that there was more bad than good in life. South Koreans having a pessimistic mentality in life suggests that life matters could easily manifest into their work domains if the individual failed to suppress it.

6.7.2 Assessment of the Bottom 5 Items in Work–Life Balance Assessment

Rank 93: LD5: I would not like to spend more time in my non-work life

LD5 belongs to the Life Demands indicator where an increase in demand in the life domain would limit the ability of an individual to meet his/her obligations in the work domain (Voydanoff, 2005b). It is important to note that LD5 was reverse-coded to a positive item and the results for this item shown correspondingly in Tables 6.8 and 6.9. LD5 has the lowest mean difference of −1.16 and the mean scores for Singapore and South Korea were 3.09 and 4.25 respectively. Tables 6.8 and 6.9 showed that close to 82% of the South Korean respondents agreed that they

Table 6.8 Breakdown of the mean scores for the bottom 5 items (Singapore)

Singapore	Mean score	Strongly disagree	Disagree	Neither agree nor disagree	Agree	Strongly agree
WD3	3.61	1	6	7	10	9
%		3.0%	18.2%	21.2%	30.3%	27.3%
LD3	3.52	2	9	3	8	11
%		6.1%	27.3%	9.1%	24.2%	33.3%
LD4	3.42	3	8	2	12	8
%		9.1%	24.2%	6.1%	36.4%	24.2%
WD4	3.48	1	3	13	11	5
%		3.0%	9.1%	39.4%	33.3%	15.2%
LD5	3.09	7	8	3	5	10
%		21.2%	24.2%	9.1%	15.2%	30.3%

Table 6.9 Breakdown of the mean scores for the bottom 5 items (South Korea)

South Korea	Mean score	Strongly disagree	Disagree	Neither agree nor disagree	Agree	Strongly agree
WD3	4.18	0	3	2	23	16
%		0.0%	6.8%	4.5%	52.3%	36.4%
LD3	4.14	1	4	5	12	22
%		2.3%	9.1%	11.4%	27.3%	50.0%
LD4	4.09	1	4	4	16	19
%		2.3%	9.1%	9.1%	36.4%	43.2%
WD4	4.16	0	5	4	14	21
%		0.0%	11.4%	9.1%	31.8%	47.7%
LD5	4.25	2	1	5	12	24
%		4.5%	2.3%	11.4%	27.3%	54.5%

would like to spend more time in their non-work life as compared to Singaporean respondents of 45%. As discussed earlier, South Koreans followed a strict hierarchical structure in their society especially in the workplace where the employees would not leave the office until their superiors have left the workplace (Bae, 2017; Olsen, 2008). They were perceived to be lazy or not hardworking enough if they left work earlier or on time (Olsen, 2008). This usually caused them to work overtime even though they have nothing to do. In addition, due to the nature of the construction industry, working overtime could be unavoidable at some point in time especially when meeting tight deadlines. Hence, it was not surprising to see South Korea having a high mean score for this item. Singapore on the other hand has a relatively balanced breakdown of the scores where the proportion of respondents who disagreed and disagreed with this statement were 45.5 and 45.4% respectively. The high percentage of disagreement for Singapore (45.4%) as compared to South Korea (6.8%) suggests that Millennials Singaporean were less willing to spend more time in non-work life. As shown in the PMR1 analysis, Singaporeans were more willing to accept challenging tasks than the South Koreans. Individuals pursuing promotion in the organization were more willing to accept challenging tasks to support their career advancement (Lingard & Francis, 2009). This could be one of the reasons why 45% of the Singaporean respondents disagreed they would like to spend more time in their life domain.

Rank 92: WD4: I am uncertain as to how my job is linked

WD4 belongs to the Work Demands indicator where an increase in WD4 would reduce the ability of an individual to meet his/her obligations in the life domain (Voydanoff, 2005b). WD4 was reverse-coded to a positive item and the results shown in Tables 6.8 and 6.9. WD4 has a mean difference of −0.67 and the mean scores for Singapore and South Korea were 3.48 and 4.16 respectively. Tables 6.6 and 6.7 showed that close to 80% of the South Korean respondents agreed that they were certain as to how their job was linked compared to the Singaporean respondents of 46%. As discussed earlier, South Korea has a high score for Hofstede's cultural dimension of Uncertainty Avoidance which means South Koreans do not tolerate uncertainties and ambiguity and would follow strict norms to minimize the occurrence of the unknown (Hofstede, 2001). Although only 46% of the Singaporean respondents agreed with this statement, only 12% of them disagreed. Similarly, the proportion of South Korean respondents that disagreed with this statement was around 12%. Because the duration between awarding the contract to the start of construction was relatively short, the project team must be assembled quickly and all team members must be certain of their roles and responsibilities prior to the start of construction. Thus, it was not unexpected that a low proportion of the respondents disagreed with this statement.

Rank 91: LD4: I do not know what is expected of me in my life

LD4 belongs to the Life Demands indicator where an increase in demand in the life domain would limit the ability of an individual to meet his/her obligations in the work domain (Voydanoff, 2005b). LD4 was reverse-coded to a positive item and the results shown in Tables 6.8 and 6.9. LD4 has a mean difference of −0.67 and the mean scores for Singapore and South Korea were 3.42 and 4.09 respectively. Based on the scores for LD4, it could be seen that close to 80% of the South Korean respondents agreed that they knew what was expected of them. On the other hand, 60% of the Singaporean respondents agreed with this statement. Based on Hofstede's cultural dimension of Long Term Orientation (LTO), both Singapore and South Korea have high scores of 72 and 100 respectively. This suggests that both Singaporeans and South Koreans were pragmatic and have long-term oriented societies with cultural qualities that support their long-term goals (Hofstede et al., 2010). The pragmatic mindset suggests that Singaporeans and South Koreans were always preparing for their future and understood what was needed for them to be a better person (Hofstede, 2001). Apart from the high LTO score, South Korea also has a high score in the Hofstede's cultural dimension of Uncertainty Avoidance where South Koreans would tend to avoid ambiguous situations in life and would follow the social norms even though these might not always work (Hofstede, 2001). With a low tolerance for ambiguity in their life, this would undoubtedly lead the South Koreans to be more aware of what was expected of them in their life domain.

Rank 90: LD3: I do not know exactly what my responsibilities are in life

LD3 belongs to the Life Demands indicator where an increase in demand in the life domain would limit the ability of an individual to meet his/her obligations in the work domain. LD3 and LD4 belonged to the same sub-indicator called Life Role Ambiguity. Life Role Ambiguity has been defined as a lack of the necessary information about the roles and responsibilities needed to perform particular life-role (Michel et al., 2010). LD3 was reverse-coded to a positive item and the results shown in Tables 6.8 and 6.9. LD3 has a mean difference score of −0.62 and the mean scores for Singapore and South Korea were 3.52 and 4.14 respectively. Tables 6.8 and 6.9 showed that LD3 has quite similar results as LD4 where close to 78% of the South Korean respondents and 58% of the Singaporean respondents agreed they knew exactly what their responsibilities were in life. Hence the assessment results for LD3 were quite similar to LD4.

Rank 89: WD3: I do not know exactly what my responsibilities are at work

WD3 belongs to the Work Demands indicator where an increase in demand in the work domain would limit the ability of an individual to meet his/her obligations in the life domain. WD3 and WD4 belonged to the sub-indicator called Work Role Ambiguity. Work Role Ambiguity has been defined as a lack of the necessary information about the roles and responsibilities needed to perform the work-role (Michel et al., 2010). WD3 was reverse-coded to a positive item and the results shown in Tables 6.8 and 6.9. WD3 has a mean difference of −0.58 (with rounding off) and the mean scores for Singapore and South Korea were 3.61 and 4.18 respectively. The results showed that 89% of the South Koreans and 58% of the Singaporeans agreed that they knew exactly what their responsibilities were at work. The assessment of WD4, LD3 and LD4 showed that the South Koreans knew exactly what their responsibilities were and what was expected of them at work or in life. On average, more than 80% of the South Korean respondents agreed with these four items. All the results concurred with the Hofstede's cultural dimension of Uncertainty Avoidance where South Korea was described as a society that avoided uncertainties and ambiguity in life and would follow social norms to reduce any unknown occurrence. The other factor to take note of relates to the nature of the construction industry. Because the duration between the award of contract to the start of construction was relatively short, the project team must be assembled quickly and all team members must be certain of their roles and responsibilities prior to the start of construction. This could be one of the reasons as to why more than half of the Singaporean and South Korean respondents agreed with this statement.

6.8 Assessment of the Work–Life Balance Indicator Between Singapore and South Korea

In this analysis, the mean differences between the Singapore scores and the South Korea scores for WLB indicator were assessed and compared. Table 6.10 shows the rankings of all the items in descending order. There were six items in this indicator and their mean differences ranged between −0.05 and −0.36. The items in the WLB indicators were based on the study conducted by Carlson, Grzywacz, and Zivnuska (2009) where the WLB of an individual was assessed through the social perspective. The items were designed to benefit from the definition given by Grzywacz and Carlson (2007: 458) which states that WLB is the "accomplishment of role-related expectations that are negotiated and shared between an individual and his or her role-related partners in the work and family/life domains". This means that instead of individuals assessing their own level of WLB, Carlson et al. (2009) used the social domain's perspective (role-related partners) to evaluate whether an individual has achieved WLB.

Table 6.10 Mean differences of Work–Life Balance indicators between Singapore and South Korea

Ranking	Indicators	Mean difference
1	WLB5	−0.053
2	WLB6	−0.129
3	WLB3	−0.174
4	WLB1	−0.197
5	WLB4	−0.235
6	WLB2	−0.356

6.8.1 Assessment of the Items in Work–Life Balance Indicator

Rank 1: WLB5: People who are close to me would say that I do a good job of balancing work and life

WLB5 belongs to the Work–Life Balance indicator where an increase in WLB would yield beneficial outcomes in both the work and life domains. WLB5 has a mean difference of −0.05 and the mean scores for Singapore and South Korea were 3.52 and 3.57 respectively. Tables 6.11 and 6.12 showed that close to 70% of the respondents in Singapore and South Korea agreed people who were close to them would say they did a good job balancing their work and life. The marginal mean difference was not unexpected because both countries belonged to a feminine society as described in Hofstede's cultural dimension of Masculinity versus Femininity. This suggests that Singaporeans and South Koreans were generous in complimenting each other when they did a good job. Another possible reason for this outcome was because Singapore and South Korea belonged to a collectivistic

Table 6.11 Breakdown of the mean scores for items in the Work–Life Balance indicators (Singapore)

Singapore	Mean score	Strongly disagree	Disagree	Neither agree nor disagree	Agree	Strongly agree
WLB5	3.52	1	5	8	14	5
%		3.0%	15.2%	24.2%	42.4%	15.2%
WLB6	3.58	0	2	12	17	2
%		0.0%	6.1%	36.4%	51.5%	6.1%
WLB3	3.67	1	1	10	17	4
%		3.0%	3.0%	30.3%	51.5%	12.1%
WLB1	3.58	2	2	9	15	5
%		6.1%	6.1%	27.3%	45.5%	15.2%
WLB4	3.70	1	1	8	20	3
%		3.0%	3.0%	24.2%	60.6%	9.1%
WLB2	3.48	3	2	7	18	3
%		9.1%	6.1%	21.2%	54.5%	9.1%

Table 6.12 Breakdown of the mean scores for items in the Work–Life Balance indicators (South Korea)

South Korea	Mean score	Strongly disagree	Disagree	Neither agree nor disagree	Agree	Strongly agree
WLB5	3.57	3	5	7	22	7
%		6.8%	11.4%	15.9%	50.0%	15.9%
WLB6	3.70	0	6	9	21	8
%		0.0%	13.6%	20.5%	47.7%	18.2%
WLB3	3.84	0	1	11	26	6
%		0.0%	2.3%	25.0%	59.1%	13.6%
WLB1	3.77	0	2	11	26	5
%		0.0%	4.5%	25.0%	59.1%	11.4%
WLB4	3.93	0	4	3	29	8
%		0.0%	9.1%	6.8%	65.9%	18.2%
WLB2	3.84	1	2	7	27	7
%		2.3%	4.5%	15.9%	61.4%	15.9%

society as mentioned in Hofstede's cultural dimension of Individualism versus Collectivism. A collectivistic society worked as a group that would look after each other to foster good relationships (Hofstede, 2001). Giving compliments was encouraged and improving the quality of life of others was deemed reciprocal. However, the compliments or answers received from the individual's role-related partners might not always be true. The collectivism dimension shared similarities with the second key principle of Confucian values where "family is the prototype of all social organizations. Harmony is found in the maintenance of everybody's face, in the sense of dignity, self-respect and prestige" (Hofstede, 2001: 354). This means that open conflicts were avoided actively in order not to hurt other's feelings, and where politeness took precedence over honest feedback where a "yes" do not necessarily means "yes". Likewise, a feminine society also followed such traits to prevent others from getting discouraged or less motivated.

Rank 2: WLB6: It is clear to me, based on feedback from co-workers, friends and family members, that I am accomplishing both my work and life responsibilities

WLB5 has a mean difference of −0.13 (with rounding up). The mean scores for Singapore and South Korea were 3.58 and 3.70 respectively. Tables 6.11 and 6.12 showed that 58% of the Singaporean respondents and 66% of the South Korean respondents agreed they were accomplishing their work and life responsibilities based on the feedback from their role-related partners. As this item was based on the social domain's perspective to evaluate individual WLB, the Hofstede's cultural dimensions of Masculinity versus Femininity and Individualism versus Collectivism could be applied to this item as well. Thus, the assessment for WLB6 was quite similar to WLB5.

Rank 3: WLB3: I do a good job of meeting the role expectations of critical people in my work and life

WLB3 has a mean difference of −0.17 and the mean scores for Singapore and South Korea were 3.67 and 3.84 respectively. Tables 6.11 and 6.12 showed that 63% of Singaporean respondents and 74% of South Korean respondents agreed they did a good job of meeting the role expectations of critical people in their work and life domain. In this item, apart from the application of Hofstede's cultural dimensions of Masculinity versus Femininity and Individualism versus Collectivism explained in WLB5, this also included a self-evaluation of the individual WLB. As both Singapore and South Korea belonged to a high Confucian value-based society, the stability of society was premised on the unequal relationships between people (father-son, superior-subordinate) as stated in the first principle of Confucian teachings. This means it was the duty of the individual to fulfill the obligations of others as a form of respect and/or obedience (Hofstede, 2001). This was also the reason why Singapore and South Korea have a high Power Distance index where society was based on conformance with a strict hierarchical order.

Rank 4: WLB1: I am able to negotiate and accomplish what is expected of me at work and in my life

WLB1 has a mean difference of −0.20 (with rounding off) and the mean scores for Singapore and South Korea were 3.58 and 3.77 respectively. Tables 6.11 and 6.12 showed that 61% of the Singaporean and South Korean respondents agreed they were able to negotiate and accomplish what was expected of them at work and in life. Apart from the application of Hofstede's cultural dimensions of Masculinity versus Femininity and Individualism versus Collectivism as explained in WLB5, this item also contained a self-evaluation of the individual WLB. Singapore and South Korea have a high score in the Hofstede's cultural dimension of Long Term Orientation (LTO). Some traits associated with countries with high LTO included: persistence, perseverance, constantly acquiring new skills, personal adaptability and adaption of traditions to new circumstances (Hofstede, 2001). Such traits would allow the Singaporeans and South Koreans to have the ability to negotiate and accomplish what was expected of them. Furthermore, as South Koreans belonged to a society that do not tolerate uncertainties and ambiguity in life, they would work hard and comply with strict social norms in order to lead a better quality of life even though it might not always be right to do so (Hofstede, 2001). As such, this could also be another possible reason why the South Koreans have a high propensity to fulfill what was expected of them.

Rank 5: WLB4: My co-workers, friends and family would say that I am meeting their expectations

WLB5 has a mean difference of −0.24 (with rounding up) and the mean scores for Singapore and South Korea were 3.70 and 3.93 respectively. Tables 6.11 and 6.12 showed that 70% of the Singaporean respondents and 84% of the South Korean respondents agreed their role-related partners said that they were meeting their expectations. As this item was based on the social domain's perspective to evaluate individual WLB, the Hofstede's cultural dimensions of Masculinity versus Femininity and Individualism versus Collectivism could also be applied to this item. Thus, the assessment for WLB4 was quite similar to WLB5.

Rank 6: WLB2: I am able to accomplish the expectations that my superior and my family have for me

WLB5 has a mean difference of −0.36 and the mean scores for Singapore and South Korea were 3.48 and 3.84 respectively. Tables 6.11 and 6.12 showed that 64% of the Singaporeans and 77% of the South Koreans agreed they were able to accomplish the expectations their superiors and families have for them. As this item was based on the social domain's perspective to evaluate individual WLB, the Hofstede's cultural dimensions of Masculinity versus Femininity and Individualism versus Collectivism could also be applied to this item. Thus, the assessment for WLB2 was quite similar to WLB5.

6.9 Assessment of the Work–Life Balance Outcomes Between Singapore and South Korea

In this analysis, the mean differences for the WLB outcomes between Singapore and South Korea consisted twelve items that were ranked in descending order as shown in Table 6.13. Two indicators served as the WLB outcomes: work outcome (WO) and life outcome (LO). The highest (top 3) and lowest (bottom 3) scores would be assessed and compared. The mean differences for the WLB outcomes ranged between −0.05 and −0.60.

Table 6.13 Mean differences of Work–Life Balance outcomes between Singapore and South Korea

Ranking	Indicators	Mean difference
1	WO4	−0.045
2	LO6	−0.152
3	WO2	−0.174
4	LO4	−0.182
5	WO5	−0.189
6	LO2	−0.265
7	WO3	−0.311
8	WO1	−0.326
9	LO3	−0.409
10	LO5	−0.417
11	LO1	−0.470
12	WO6	−0.583

6.9.1 Assessment of the Top 3 Items in Work–Life Balance Outcomes

Rank 1: WO4: I really feel as if my organization's problem is my own

WO4 belongs to the Work Outcomes indicator where the achievement of WLB would yield positive work outcomes to the individual. WO4 has a mean difference of −0.05 and the mean scores for Singapore and South Korea were 3.27 and 3.32 respectively. Tables 6.14 and 6.15 showed that 50% (on average) of the Singaporean and South Korean respondents agreed they felt as if their organization's problems were their own. Only 23% (on average) of the respondents disagreed. The marginal mean difference between these two countries was not surprising because they both belonged to a collectivistic society. The Hofstede's cultural dimension of Individualism versus Collectivism suggested that a collectivistic society would work together as a group; taking responsibility as a group even though it might not be its fault and would treat each other as a member of their family. This concurred with the Confucian second principle where "a person is not primarily an individual; rather, he or she is a member of a family" (Hofstede, 2001: 354).

Table 6.14 Breakdown of the mean scores for the top 3 items in Work–Life Balance Outcomes (Singapore)

Singapore	Mean score	Strongly disagree	Disagree	Neither agree nor disagree	Agree	Strongly agree
WO4	3.27	3	5	9	12	4
%		9.1%	15.2%	27.3%	36.4%	12.1%
LO6	3.76	1	2	8	15	7
%		3.0%	6.1%	24.2%	45.5%	21.2%
WO2	3.39	1	5	9	16	2
%		3.0%	15.2%	27.3%	48.5%	6.1%

Table 6.15 Breakdown of the mean scores for the top 3 items in Work–Life Balance outcomes (South Korea)

South Korea	Mean score	Strongly disagree	Disagree	Neither agree nor disagree	Agree	Strongly agree
WO4	3.32	2	8	12	18	4
%		4.5%	18.2%	27.3%	40.9%	9.1%
LO6	3.91	1	2	8	22	11
%		2.3%	4.5%	18.2%	50.0%	25.0%
WO2	3.57	2	4	11	21	6
%		4.5%	9.1%	25.0%	47.7%	13.6%

Rank 2: LO6: For the most part, I am proud of whom I am and the life I lead

LO6 belongs to the Life Outcomes indicator where the achievement of WLB would yield positive life outcomes to the individual. LO6 has a mean difference of −0.15 (with rounding off) and the mean scores for Singapore and South Korea were 3.76 and 3.91 respectively. Tables 6.14 and 6.15 showed that 68% of the Singaporean respondents and 75% of the South Korean respondents agreed they were proud of themselves and the life they led. By applying the Hofstede's cultural dimension of Uncertainty Avoidance, it appears that both societies in Singapore and South Korea have a different perspective to explain this phenomenon. Singapore has a low Uncertainty Avoidance index that indicates Singaporeans accepted uncertainties and ambiguity in life. A low Uncertainty Avoidance society also exhibits a more carefree or stress-free attitude and has the ability to adapt to changes. This could be a possible reason why 68% of the Singaporeans respondents were content with their life. South Korea on the other hand has a high Uncertainty Avoidance index that indicates South Koreans avoided uncertainties and ambiguity in life and would follow strict norms to lead a better life even though these might not always be the best solution. This means that for the South Koreans to be content with their life, they have to maintain the strict norms and behaviors instilled in the South Korea society. Thus, it was very likely that 7% of the South Korean respondents who disagreed with this item were those who do not follow the norms and behaviors in the society.

Rank 3: WO2: All in all, I am satisfied in my job

WO2 has a mean difference of −0.17 (with rounding off) and the mean scores for Singapore and South Korea were 3.39 and 3.57 respectively. Tables 6.14 and 6.15 showed that 55% of the Singaporean respondents and 61% of the South Korean respondents agreed they were satisfied with their jobs. The construction industry has always been perceived to be dirty, dangerous and demanding (Seah, 2013). Work overload and working beyond normal work hours were some of the unavoidable situations when projects experienced tight deadlines. However, the result showed that more than 50% of the Singaporean and South Koreans were satisfied with their jobs. One possibility could be that both countries belonged to a collectivistic society. Hofstede (2001) mentioned that a collectivistic society has a "We" mentality and would look after one another in exchange for loyalty. This means the group would work towards collective goals and success or failure shared among the group and no individual was to be blamed (Hofstede et al., 2010). This is exceptionally important in construction projects involving numerous professional disciplines that must be assembled quickly prior to the start of construction. A well-coordinated workflow among the project team members throughout the entire construction period would help the individuals accomplished their task effectively and efficiently. This would eventually result in them feeling more satisfied with their work.

6.9.2 Assessment of the Bottom 3 Items in Work–Life Balance Outcomes

Rank 12: WO6: I will not give up on this job easily

WO6 has a mean difference of −0.58 and the mean scores for Singapore and South Korea were 3.39 and 3.98 respectively. Tables 6.16 and 6.17 showed that 62% of the Singaporean respondents and 66% of the South Korean respondents agreed they would not give up on their job easily. Low and Shi (2002) mentioned that Singapore has a low Uncertainty Avoidance index (UAI) and one of the consequences of low UAI was the high tendency for turnover. People living in a low UAI society were willing to take risks and accept uncertainties and ambiguities in life. However, the results do not reflect what Low and Shi (2002) have observed. One possible reason could be Singapore has a high score in Long Term Orientation dimension. This suggests Singaporeans have a more pragmatic, future-oriented thinking and would persevere and work hard in order to acquire their goals. This also means the respondents might have considered their career prospects and have already set targeted goals prior to participating in this survey. South Korea on the other hand has a high UAI. One of the consequences associated with this was the

Table 6.16 Breakdown of the mean scores for the bottom 3 items in Work–Life Balance outcomes (Singapore)

Singapore	Mean score	Strongly disagree	Disagree	Neither agree nor disagree	Agree	Strongly agree
LO5	3.52	2	4	5	19	3
%		6.1%	12.1%	15.2%	57.6%	9.1%
LO1	3.48	1	6	7	14	5
%		3.0%	18.2%	21.2%	42.4%	15.2%
WO6	3.39	2	5	9	12	5
%		6.1%	15.2%	27.3%	36.4%	15.2%

Table 6.17 Breakdown of the mean scores for the bottom 3 items in Work–Life Balance outcomes (South Korea)

South Korea	Mean score	Strongly disagree	Disagree	Neither agree nor disagree	Agree	Strongly agree
LO5	3.93	2	3	3	24	12
%		4.5%	6.8%	6.8%	54.5%	27.3%
LO1	3.95	1	5	7	13	18
%		2.3%	11.4%	15.9%	29.5%	40.9%
WO6	3.98	0	2	13	13	16
%		0.0%	4.5%	29.5%	29.5%	36.4%

low tendency for labor turnover (Low & Shi, 2002). Hofstede (2001) mentioned that high UAI societies tended to avoid uncertainties and ambiguities in life and would follow strict social norms. Hence, it was very likely that this dimension was the main reason for the South Korean respondents not to give up their jobs.

Rank 11: LO1: The conditions of my family life are excellent

LO1 has a mean difference of −0.47 and the mean scores for Singapore and South Korea were 3.48 and 3.95 respectively. Tables 6.16 and 6.17 showed that 58% of the Singaporean respondents and 70% of the South Korean respondents agreed the conditions of their family life were excellent. This result was not unexpected because both countries belonged to a Confucian society where filial piety, family relationships and maintaining strong ties between generations were inculcated into the people since young (Yang & Rosenblatt, 2008). Furthermore, the first principle of Confucian teachings mentioned that stability of society was based on unbalanced relationships among people: brother-younger brother, husband-wife and father-son (Hofstede, 2001). These relationships were based on mutual obligations and were seen as reciprocal relationships (Hofstede, 2001). Hence, family life could be seen as excellent when every family members were committed to their roles and in providing unconditional love to each other.

Rank 10: LO5: In general, I feel confident and positive about myself

LO5 has a mean difference of −0.42 (with rounding off) and the mean scores for Singapore and South Korea were 3.52 and 3.93 respectively. Tables 6.16 and 6.17 showed that 67% of the Singaporean respondents and 82% of the South Korean respondents agreed they were confident and positive about themselves. By applying the Hofstede's cultural dimension of Uncertainty Avoidance here, it seems that the societies in both Singapore and South Korea have different perspectives to explain this phenomenon. As Singapore belonged to a low UAI society, Singaporeans felt less threatened by uncertainties and ambiguities and exhibited a more positive attitude in life (Hofstede, 2001). This could be one of the reasons why a large proportion of the Singaporeans agreed with this item. South Korea on the other hand has a high UAI indicating that the South Koreans would avoid uncertainties and ambiguities in life at all costs and to follow strict social norms in order to lead a better life even though this might not always be the best solution (Hofstede, 2001). The South Koreans would adopt other channels to stay positive in life by maintaining the strict norms and behaviors instilled in their society. Thus, it was very likely that 82% of the South Korean respondents who agreed with this item were those who followed the norms and behaviors in the society.

6.10 Inferential Statistics

The SmartPLS software was used to perform the multivariate analysis for the conceptual framework. SmartPLS is a Structural Equation Modelling (SEM) tool that simultaneously examines the relationships between the latent variables as well as among the latent variables and its observed variable (Hair, Hult, Ringle, & Sarstedt, 2017).

Before abstracting the scores from SPSS to SmartPLS, this study has combined items which belonged to the same sub-indicator to form a multi-item scale. For example, WD1 and WD2 belonged to the same sub-indicator called Work Role Conflict. The scores from both indicators were combined using simple summation as the weighting for each item was assumed to be similar. The logic behind combining similar items to measure the sub-indicator was that such measures were deemed to be more accurate and reliable as compared to a single-item construct (Hair et al., 2017). Furthermore, this also helped to reduce the complexity of the analysis by restricting the model to two layers (observed variables and latent variables) (Hair et al., 2017). Table 6.18 shows the items that were grouped to represent the sub-indicators. The sub-indicator codes shown in Table 6.18 would be used for this analysis.

The Proposed Measures indicator was excluded from the analysis because the indicator does not alter the outcome of the actual WLC and WLE that an individual was experiencing. These items were suggested recommendations proposed for this study that could enable organizations or parties concerned to understand which measure was more important to help individuals reduce their WLC and improve WLE. After computing the composite scores for all the sub-indicators, the data was imported into the SmartPLS software. A path model was created by connecting all the relevant sub-indicators (observed variables) to their latent variables. For example, WDemand1, WDemand2 and WDemand3 belonged to the concept called Work Demands. The three observed variables would then be dragged into the model and the construct for Work Demands (WD) would automatically be formed. Figure 6.1 shows the actual path model created using SmartPLS.

This study has adopted a reflective measurement model to analyze the conceptual framework. A reflective model means that the causality was from the latent variables to the observed variables. This was common for surveys that used Likert scales where observed variables were seen as the reflection of the latent variables and shared a common theme. For example, consider the construct measuring Work–Life Conflict, where an individual's level of WLC would influence what he or she answered the items associated with WLC. Another characteristic associated with the reflective measurement model was that the observed indicators were interchangeable meaning that any removal of items related to the latent variable would not change the outcome of the construct (Hair et al., 2017). The reflective model would also be tested empirically by using measures such as Composite reliability, Average Variance Extracted and Outer Loadings. All measurements would be assessed using

Table 6.18 Items designed to represent the sub-indicators

No	Item(s)	Sub-indicator name	Sub-indicator code	No	Item(s)	Sub-indicator name	Sub-indicator code
1	WD1	Work Role Conflict	WDemand1	24	WLCL1	Time-based Life to Work Conflict	WLC4
	WD2				WLCL2		
2	WD3	Work Role Ambiguity	WDemand2		WLCL3		
	WD4			25	WLCL4	Strain-based Life to Work Conflict	WLC5
3	WD5	Work Involvement	WDemand3		WLCL5		
	WD6				WLCL6		
4	BSD1	Work Overload	BoundarySD1	26	WLCL7	Behavior-based Life to Work Conflict	WLC6
	BSD2				WLCL8		
5	BSD3	Bring Work to Home	BoundarySD2		WLCL9		
	BSD4			27	WLEW1	Work to Life Development	WLE1
6	BSD5	Life Activities at Work	BoundarySD3		WLEW2		
	BSD6				WLEW3		
7	BSD7	Life Interruptions and Distraction	BoundarySD4	28	WLEW4	Work to Life Affect	WLE2
	BSD8				WLEW5		
8	LD1	Life Role Conflict	LDemand1		WLEW6		
	LD2			29	WLEW7	Work to Life Capital	WLE3
9	LD3	Life Role Ambiguity	LDemand2		WLEW8		
	LD4				WLEW9		
10	LD5	Life Involvement	LDemand3	30	WLEL1	Life to Work Development	WLE4
	LD6				WLEL2		
11	WR1	Job Salience	WResource1		WLEL3		
	WR2			31	WLEL4	Life to Work Affect	WLE5
12	WR3	Developmental Experiences	WResource2		WLEL5		
	WR4				WLEL6		

(continued)

Table 6.18 (continued)

No	Item(s)	Sub-indicator name	Sub-indicator code
13	WR5	Autonomy	WResource3
	WR6		
14	BSR1	Flexible Work Schedule	BoundarySR1
	BSR2		
15	BSR3	Time-off for Non-Work Purposes	BoundarySR2
	BSR4		
16	BSR5	Participate in Workshops for Personal Development	BoundarySR3
	BSR6		
17	BSR7	Assigned Time for Work Purposes	BoundarySR4
	BSR8		
18	LR1	Life Salience	LResource1
	LR2		
19	LR3	Life Harmony	LResource2
	LR4		
20	LR5	Life Rewards	LResource3
	LR6		
21	WLCW1	Time-based Work to Life Conflict	WLC1
	WLCW2		
	WLCW3		
22	WLCW4	Strain-based Work to Life Conflict	WLC2
	WLCW5		
	WLCW6		

No	Item(s)	Sub-indicator name	Sub-indicator code
32	WLEL7	Life to Work Efficiency	WLE6
	WLEL8		
	WLEL9		
33	CS1	Colleagues and Employer Support	CS1
34	CS2	Families and Friends Support	CS2
35	CS3	Leisure Activities	CS3
36	CS4	Cognitive Management	CS4
37	CS5	Behavioral Management	CS5
38	WLB1	Work–Life Balance	WorkLifeBalance
	WLB2		
	WLB3		
	WLB4		
	WLB5		
	WLB6		
39	WO1	Job Satisfaction	OutcomeWork1
	WO2		
40	WO3	Organizational Commitment	OutcomeWork2
	WO4		
41	WO5	Intention to Turnover	OutcomeWork3
	WO6		
42	LO1	Family Satisfaction	OutcomeLife1
	LO2		

(continued)

Table 6.18 (continued)

No	Item(s)	Sub-indicator name	Sub-indicator code	No	Item(s)	Sub-indicator name	Sub-indicator code
23	WLCW7	Behavior-based Work to Life Conflict	WLC3	43	LO3	Life Satisfaction	OutcomeLife2
	WLCW8				LO4		
	WLCW9			44	LO5	Psychological Well-Being	OutcomeLife3
					LO6		

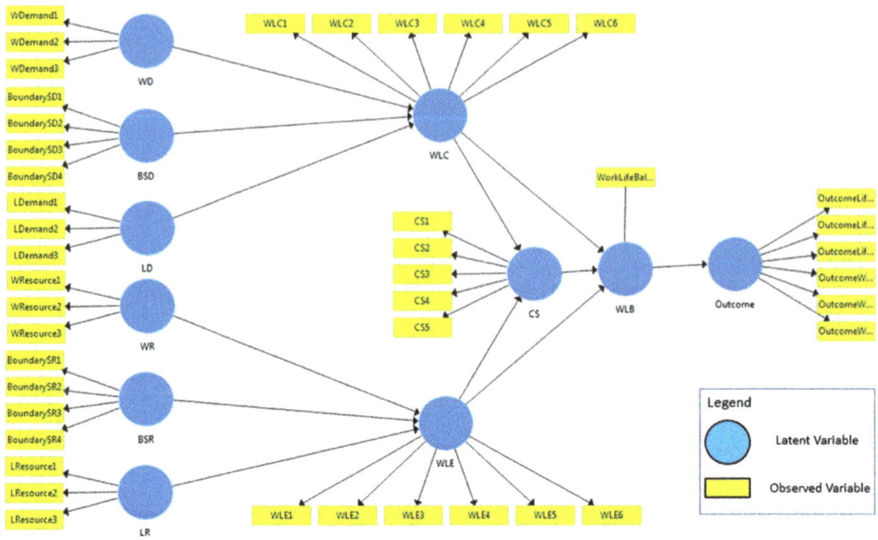

Fig. 6.1 Path model of the conceptual framework created by SmartPLS

SmartPLS and the evaluation of the reflective model would be based primarily on the research presented by Hair et al. (2017).

There are two parts to this analysis: assessment of reflective measurement model and the assessment of the structural model. All the methods related to the assessment for reflective measurement model would be conducted to determine how well the theory fits the data. The results would be reviewed and evaluated using the rule of thumb stated by Hair et al. (2017). Once after the reliability and validity of the construct have been established, assessment of structural model would be addressed. In this part, a series of steps was conducted to examine the model's predictive capabilities as well as the relationships between the latent variables (Hair et al., 2017).

The following statistical criteria were extracted from the results that examined the reflective measurement model: Indicator Reliability (Outer Loadings), Convergent Validity (Average Variance Extracted (AVE)), Internal Consistency (Composite Reliability) and Discriminant Validity. The key to this measurement was to ensure that the entire statistics criteria met the targeted number as stipulated by Hair et al. (2017). If the model did not meet one of the criteria, deletion of the indicator was needed to ensure that the construct was valid and reliable. After the construct's validity and reliability have been established, the next step was to assess the structural model results. The assessment of the structural model allowed the research team to determine the model's capabilities in predicting the relationships between constructs.

6.11 Assessment of Reflective Measurement Model on Singapore Data

6.11.1 Assessing the Statistical Criterion of Reflective Measurement Model Before Deletion

The first step in analyzing the results was to run the PLS algorithm and the SmartPLS results report would automatically be generated. Figure 6.2 presents the outer loadings between the construct and observed variables as well as the path coefficients between the constructs. Note that only the outer loading would be required to assess the reflective measurement model; path coefficients would only be needed during the assessment of the structural model. The outer loadings were part of the indicator reliability test and represented how much the associated observed variables had in common with the construct (latent variables) (Hair et al., 2017).

The common rule of thumb is that the outer loadings of each observed variables must be above 0.708 (Hair et al., 2017). The reason is that by squaring the outer loading, it shows how much variation in an item is explained by the construct. Squaring 0.708 will give a score of 0.5 and this means the latent variable should explain at least 50% of each of its indicator's variance. Any indicators below the threshold must be eliminated as it is deemed unreliable in the measurement. However, Hair et al. (2017) mentioned that many researchers have obtained weaker outer loadings of below 0.700 especially in newly developed scales. In this context, it was found that instead of automatically deleting the indicators, the effects of the removal of the indicator must be examined first. In general, outer loadings ranging from 0.40 to 0.70 should only be considered for deletion if deleting will lead to an increase in the AVE and Composite Reliability above the threshold value (Hair et al., 2017). In contrast, outer loadings of below 0.40 should always be eliminated from the construct. Table 6.19 shows the results for the outer loadings for Singapore and observed variables with outer loadings less than 0.40 were highlighted in red. BoundarySD3 (0.126), LDemand1 (−0.085), BoundarySR1 (−0.195), and OutcomeWork3 (0.045) were eliminated from the construct. WDemand1 (0.671), WDemand2 (0.695), BoundarySD1 (0.660), BoundarySD2 (0.615), WResource1 (0.671), WLC5 (0.668), CS3 (0.438), CS5 (0.660), OutcomeLife1 (0.656) and OutcomeLife2 (0.657) were retained and would be further assessed to determine their validity and reliability. Deletion is required if retaining the observed variables (highlighted in green) would result in the criterion not meeting its targeted value. Note that even though the outer loadings of the single-item variable WorkLifeBalance was 1.00, this could not be used to interpret as evidence that the construct of WLB was reliable and would not be reported with other measures of validity and reliability (Hair et al., 2017).

Internal consistency is to test the inter-correlations between variables of the interchangeable reflective observed indicators. Composite Reliability was used to test the reliability for internal consistency instead of the Cronbach's Alpha. The

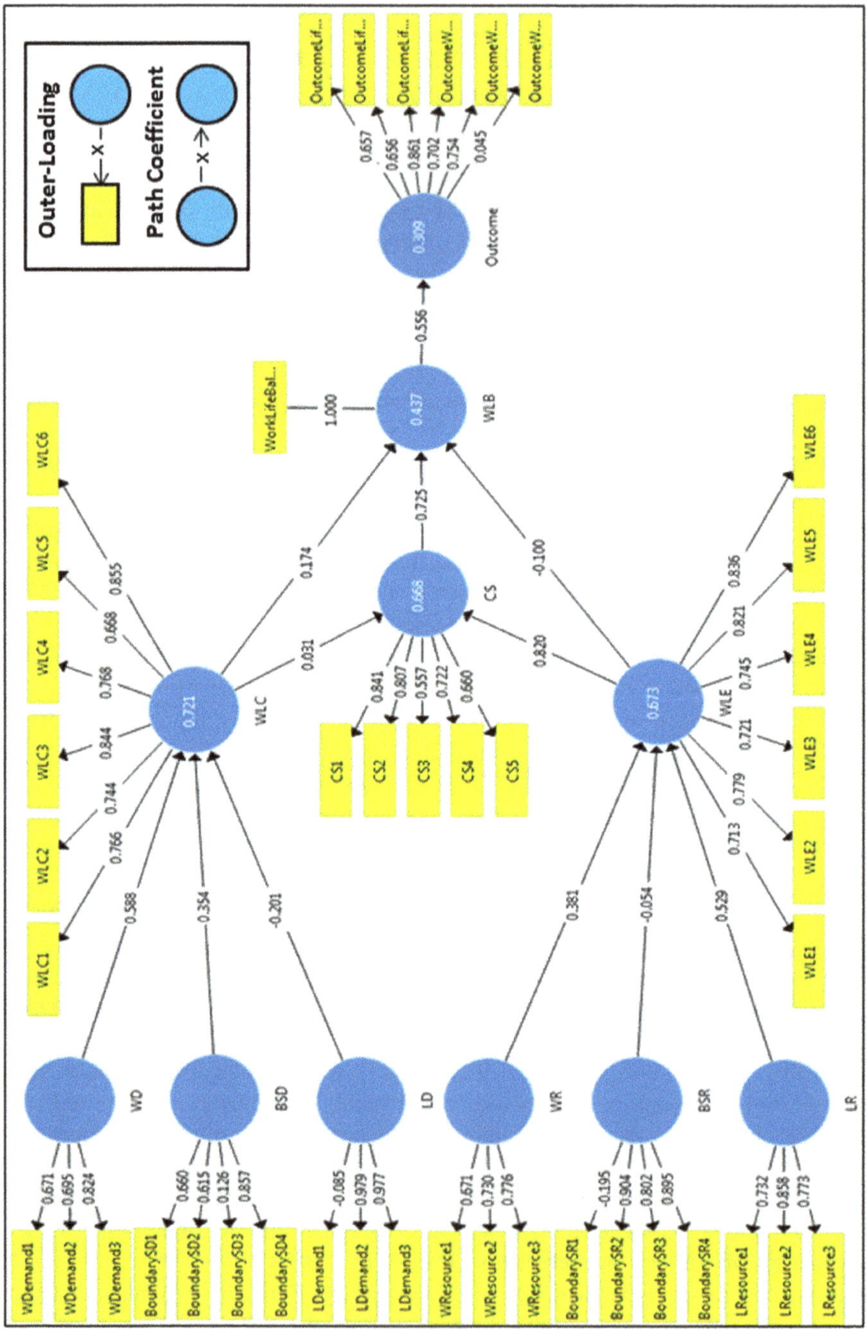

Fig. 6.2 Outer loadings and path coefficients of the reflective measurement model for Singapore result (before deletion)

Table 6.19 Results for the outer loadings before deletion (Singapore)

Outer Loadings											
	WD	BSD	LD	WR	BSR	LR	WLC	WLE	CS	WLB	Outcome
WDemand1	0.671										
WDemand2	0.694										
WDemand3	0.824										
BoundarySD1		0.660									
BoundarySD2		0.615									
BoundarySD3		0.126									
BoundarySD4		0.857									
LDemand1			-0.085								
LDemand2			0.979								
LDemand3			0.977								
WResource1				0.671							
WResource2				0.730							
WResource3				0.776							
BoundarySR1					-0.195						
BoundarySR2					0.904						
BoundarySR3					0.802						
BoundarySR4					0.895						
LResource1						0.732					
LResource2						0.858					
LResource3						0.773					
WLC1							0.766				
WLC2							0.744				
WLC3							0.844				
WLC4							0.768				
WLC5							0.668				
WLC6							0.855				
WLE1								0.713			
WLE2								0.779			
WLE3								0.721			
WLE4								0.745			
WLE5								0.821			
WLE6								0.836			
CS1									0.841		
CS2									0.807		
CS3									0.557		
CS4									0.722		
CS5									0.660		
WorkLifeBalance										1.000	
OutcomeWork1											0.702
OutcomeWork2											0.754
OutcomeWork3											0.045
OutcomeLife1											0.657
OutcomeLife2											0.656
OutcomeLife3											0.861

reason was because the Cronbach's Alpha tends to underestimate the internal consistency as it fails to take into account the different outer loadings of each indicator and assumed it to have equal loadings (Hair et al., 2017). Figure 6.3 presents the Composite Reliability of the test results before deletion. A Composite Reliability of 0.70 and above is deemed to be satisfactory. The results showed that all the constructs, except for BSD, have a Composite Reliability of 0.68. Hence, BSD was further assessed to determine whether the deletion of the outer loading would lead to an increase in the Composite Reliability.

Hair et al. (2017: 112) defined Convergent Validity as the "extent to which a measure correlates positively with alternative observed variables of the same construct". A common measure in establishing Convergent Validity is by using the statistical criterion called Average Variance Extracted (AVE). The AVE is defined as the "grand mean value of the squared loadings of the indicators associated with the construct" (Hair et al., 2017: 114). This means the AVE is calculated by summing the squared outer loadings of all the observed variables in the same construct divided by the number of observed variables. The rule of thumb is that the AVE must be at least 0.50 in order for the construct to be valid. The graph depicted in Fig. 6.4 showed that most of the constructs have an AVE value of 0.50 and above except for the BSD and Outcome constructs which have AVE values of 0.39 and 0.45 respectively.

Hair et al. (2017: 115) defined discriminate validity as "the extent to which a construct is truly distinct from other constructs". There are two approaches to assess the discriminate validity: Cross-Loadings and the Fornell-Larcker criterion. Hair et al. (2017) mentioned that an indicator's outer loading must always be greater than any of the cross-loadings of other constructs. As shown in Table 6.20, the numbers indicated in red are the loadings from the constructs while the numbers in black are the cross-loadings from other constructs. There were indicators (highlighted in green) in BSR and CS that have cross-loadings that exceeded the loading

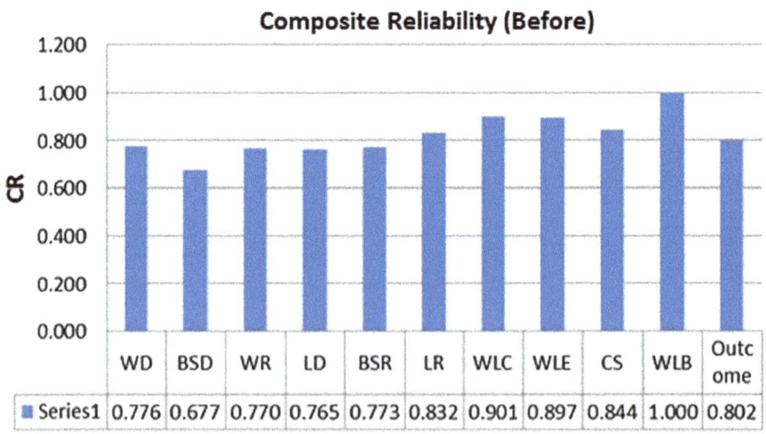

Fig. 6.3 Composite reliability results of Singapore (before deletion)

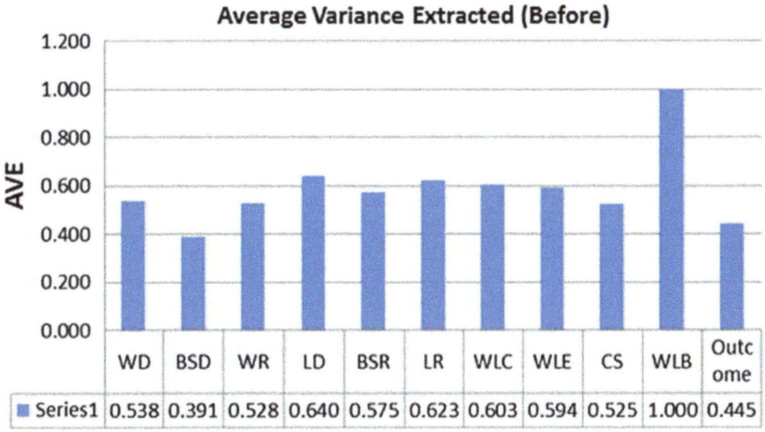

Fig. 6.4 Average variance extracted results of Singapore (before deletion)

of the constructs (BoundarySD3 in BSR; OutcomeLife3, OutcomeWork1, WResource1 and WorkLifeBalance in CS). The cross-loadings would be re-evaluated after the removal of the indicator from the construct. The loadings from its own construct highlighted in grey were not assessed as these would be removed from the model due to the poor loading. The second approach to assess discriminant validity is by using the Fornell-Larcker criterion. This criterion compares the square root of the AVE values against the correlations of other latent variables. In order to establish the discriminant validity, the value of the square root of AVE must be greater than the highest correlation of other constructs. For example, BSR has a square root AVE of 0.788 and it could be seen that the correlation values of other constructs within the column have values lesser than 0.788. This means the discriminant validity was established for BSR. However, as shown in Table 6.21, values highlighted in red are values higher than the square root AVE of the construct: WD (0.768) > BSD (0.626), WLC (0.751) > BSD (0.626), and WLC (0.799) > WD (0.733). To establish the discriminant validity of the entire model, the indicator from the construct has to be removed to increase the square root AVE.

6.11.2 Assessing the Statistical Criterion of Reflective Measurement Model After Deletion

After assessing the statistical criterion, some of the observed variables were removed to meet the requirements stated by the criterion. Figure 6.5 shows the outer loadings model and the path coefficients of the model after deletion. Seven observed variables in total were removed in order to preserve the validity and

Table 6.20 Cross-loadings results (before deletion)

Cross-Loadings											
	BSD	BSR	CS	LD	LR	OC	WD	WLB	WLC	WLE	WR
BoundarySD1	0.660	0.307	-0.372	0.002	-0.012	-0.282	0.439	-0.310	0.343	-0.289	-0.175
BoundarySD2	0.615	0.599	0.095	0.477	0.147	-0.024	0.426	0.025	0.375	0.058	-0.006
BoundarySD3	0.126	0.802	-0.150	0.318	-0.355	-0.520	0.047	-0.027	-0.089	-0.363	-0.282
BoundarySD4	0.857	0.018	-0.061	0.198	0.109	-0.013	0.082	0.252	0.165	-0.042	-0.124
BoundarySR1	-0.407	-0.195	0.212	-0.103	-0.255	0.003	-0.365	0.002	-0.340	0.058	0.009
BoundarySR2	0.251	0.904	-0.160	0.324	-0.339	-0.117	0.043	-0.134	0.039	-0.346	-0.345
BoundarySR3	0.126	0.802	-0.150	0.318	-0.355	-0.052	0.047	-0.027	-0.089	-0.363	-0.282
BoundarySR4	0.222	0.895	-0.084	0.341	-0.355	-0.014	-0.003	-0.121	0.682	-0.249	-0.185
CS1	-0.033	-0.068	0.841	0.138	0.498	0.332	-0.087	0.546	-0.044	0.420	0.478
CS2	-0.069	-0.154	0.807	0.190	0.251	0.459	-0.073	0.403	-0.014	0.635	0.457
CS3	-0.135	-0.267	0.557	0.095	0.667	0.393	-0.050	0.394	-0.185	0.553	0.449
CS4	0.092	0.078	0.722	0.355	0.044	0.506	0.014	0.431	0.128	0.445	0.294
CS5	-0.278	-0.210	0.660	0.067	0.126	0.365	-0.200	0.502	-0.070	0.487	0.276
LDemand1	-0.141	0.068	0.018	-0.085	-0.121	-0.189	-0.173	-0.187	-0.016	0.054	-0.221
LDemand2	0.255	0.382	0.246	0.979	0.182	-0.076	0.257	0.044	0.075	0.293	0.283
LDemand3	0.241	0.386	0.197	0.977	0.184	-0.085	0.299	-0.063	0.058	0.251	0.294
LResource1	0.045	-0.298	0.179	-0.035	0.732	0.169	-0.010	0.290	0.046	0.470	0.213
LResource2	0.194	-0.236	0.399	0.203	0.858	0.299	0.311	0.316	0.134	0.605	0.454
LResource3	0.104	-0.365	0.456	0.254	0.773	0.151	0.119	0.100	0.021	0.667	0.520
OutcomeLife1	-0.165	0.024	0.338	-0.258	-0.120	0.657	-0.293	0.335	-0.036	0.231	0.140
OutcomeLife2	-0.094	0.183	0.303	-0.001	-0.010	0.656	0.083	0.325	0.087	0.075	0.135
OutcomeLife3	-0.016	-0.092	0.658	-0.017	0.371	0.861	-0.100	0.626	0.038	0.409	0.321
OutcomeWork1	-0.125	-0.249	0.579	-0.007	0.309	0.702	-0.060	0.298	-0.081	0.602	0.429
OutcomeWork2	-0.043	-0.176	0.524	0.076	0.267	0.754	0.081	0.226	0.042	0.483	0.547
OutcomeWork3	-0.144	-0.074	-0.014	-0.146	-0.414	0.045	-0.098	-0.024	0.083	-0.089	0.058
WDemand1	0.534	0.286	0.019	0.327	-0.008	-0.034	0.671	0.014	0.470	-0.046	-0.034
WDemand2	0.435	-0.187	-0.156	0.148	0.225	-0.142	0.695	-0.026	0.545	-0.102	0.029
WDemand3	0.296	0.067	-0.095	0.210	0.176	-0.428	0.824	0.222	0.071	-0.044	0.057
WLC1	0.513	-0.081	-0.091	-0.105	0.198	0.114	0.527	0.304	0.766	0.020	0.125
WLC2	0.601	-0.015	0.029	-0.052	-0.080	0.119	0.526	0.174	0.744	-0.053	-0.142
WLC3	0.411	-0.062	-0.041	0.168	0.112	0.007	0.082	-0.172	0.844	-0.080	-0.012
WLC4	0.507	0.165	-0.084	0.113	-0.050	-0.178	0.579	0.277	0.768	-0.172	-0.099
WLC5	0.397	0.230	0.240	0.353	0.194	0.183	0.445	-0.099	0.668	0.122	0.245
WLC6	0.607	-0.009	-0.194	-0.052	0.047	-0.110	0.072	0.250	0.855	-0.242	-0.209
WLE1	-0.071	-0.296	0.448	0.289	0.431	0.335	-0.001	0.375	-0.071	0.713	0.638
WLE2	-0.129	-0.315	0.503	0.229	0.435	0.347	-0.128	0.257	-0.219	0.779	0.526
WLE3	-0.180	-0.323	0.387	0.162	0.440	0.301	-0.204	0.438	-0.091	0.721	0.444
WLE4	0.119	-0.340	0.515	0.171	0.443	0.370	0.068	0.402	0.031	0.745	0.491
WLE5	0.094	-0.185	0.387	0.240	0.317	0.530	0.111	0.402	0.092	0.821	0.555
WLE6	-0.212	-0.303	0.374	0.157	0.530	0.483	-0.291	0.455	-0.270	0.836	0.455
WResource1	-0.053	-0.315	0.596	0.182	0.474	0.297	0.003	0.216	-0.040	0.599	0.671
WResource2	-0.176	-0.234	0.249	0.131	0.311	0.443	-0.014	0.238	0.029	0.401	0.730
WResource3	-0.043	-0.096	0.234	0.399	0.292	0.228	0.089	0.230	-0.081	0.395	0.776
WorkLifeBalance	0.084	-0.103	0.634	0.020	0.286	0.556	-0.017	1.000	0.145	0.474	0.315

reliability of the model. Table 6.22 shows the results for the outer loadings after deletion. It could be seen that WDemand1, BoundarySD2, BoundarySD3, LDemand1, BoundarySR1, CS3 and OutcomeWork3 were removed. It could also be seen that indicator reliability was established as the outer loadings of less than 0.40 were removed: BoundarySD3 (0.126), LDemand1 (−0.085), BoundarySR1 (−0.195) and OutcomeWork3 (0.045). As mentioned earlier, outer loadings ranging from 0.40 to 0.70 would only be kept if removing them leads to an increase in the

Table 6.21 Fornell-Larker criterion results (before deletion)

Discriminant Validity (Fornell-Larcker Criterion)											
	BSD	BSR	CS	LD	LR	OC	WD	WLB	WLC	WLE	WR
BSD	0.626										
BSR	0.252	0.758									
CS	-0.115	-0.171	0.725								
LD	0.269	0.380	0.225	0.800							
LR	0.151	-0.382	0.456	0.200	0.789						
Outcome	-0.103	-0.071	0.683	-0.057	0.262	0.667					
WD	0.768	0.062	-0.112	0.300	0.190	-0.097	0.733				
WLB	0.084	-0.103	0.634	0.020	0.286	0.556	-0.017	1.000			
WLC	0.751	0.021	-0.055	0.070	0.085	0.017	0.799	0.145	0.777		
WLE	-0.071	-0.378	0.538	0.271	0.749	0.518	-0.085	0.474	-0.104	0.771	
WR	-0.119	-0.320	0.549	0.319	0.523	0.445	0.032	0.315	-0.044	0.675	0.727

statistical criterion's reliability and validity. In addition, Figs. 6.6 and 6.7 show that the removal of the following indicators have resulted in the measurement model meeting the Composite Reliability threshold of >0.70 and the AVE threshold of >0.50. This also means that the internal consistency reliability and convergent validity were established for the model.

Three indicators were removed from the construct because the discriminant validity was not established. As shown in Table 6.20, the cross-loadings for BoundarySD3 and CS3 were higher than some of the cross-loading(s) of other construct(s). Therefore, CS3 was removed in order to establish the discriminant validity while BoundarySD3 has already been eliminated due to poor outer loading (0.126). Table 6.23 shows the results of the cross-loadings after deletion. The next approach to measure discriminant validity is the use of the Fornell-Larcker criterion. As shown in Table 6.24, two constructs (BSD and WD) were unable to establish the discriminant validity for the Fornell-Larcker criterion. The WD construct has an initial squared AVE of 0.733 which was lower than the correlation of the WLC construct (0.799). The WDemand1 indicator was removed from the construct; resulting in the square root AVE for the WD construct (0.812) greater than the correlation of other constructs. Next, the BSD construct has an initial square root AVE of 0.626 which was lower than the correlation of WD (0.768) and WLC (0.751). However, the removal of BoundarySD3 does not meet the criterion because the correlation of WLC (0.735) was still greater than the square root AVE of the BSD construct (0.723). Hence, BoundarySD2 (the next lowest loading) was

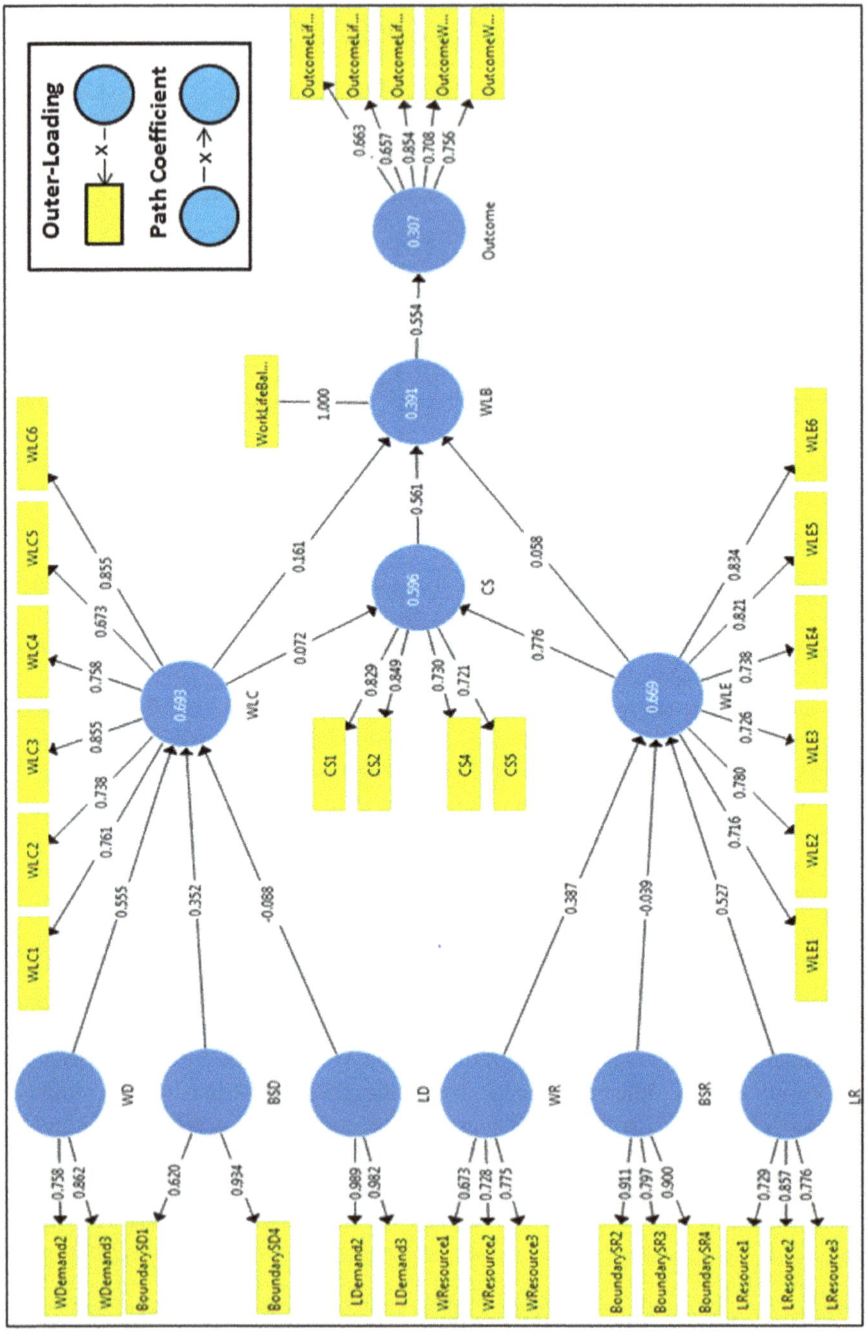

Fig. 6.5 Results of outer loadings and path coefficients of the reflective measurement model for Singapore (after deletion)

Table 6.22 Results for the outer loadings after deletion (Singapore)

Outer loadings											
	WD	BSD	LD	WR	BSR	LR	WLC	WLE	CS	WLB	Outcome
WDemand2	0.758										
WDemand3	0.862										
BoundarySD1		0.620									
BoundarySD4		0.934									
LDemand2			0.989								
LDemand3			0.982								
WResource1				0.673							
WResource2				0.728							
WResource3				0.775							
BoundarySR2					0.911						
BoundarySR3					0.797						
BoundarySR4					0.900						
LResource1						0.729					
LResource2						0.857					
LResource3						0.776					
WLC1							0.761				
WLC2							0.738				
WLC3							0.855				
WLC4							0.758				
WLC5							0.673				
WLC6							0.855				
WLE1								0.716			
WLE2								0.780			

(continued)

Table 6.22 (continued)

Outer loadings

	WD	BSD	LD	WR	BSR	LR	WLC	WLE	CS	WLB	Outcome
WLE3								0.726			
WLE4								0.738			
WLE5								0.821			
WLE6								0.834			
CS1									0.829		
CS2									0.849		
CS4									0.730		
CS5									0.721		
WorkLifeBalance										1.000	
OutcomeWork1											0.708
OutcomeWork2											0.756
OutcomeLife1											0.663
OutcomeLife2											0.657
OutcomeLife3											0.854

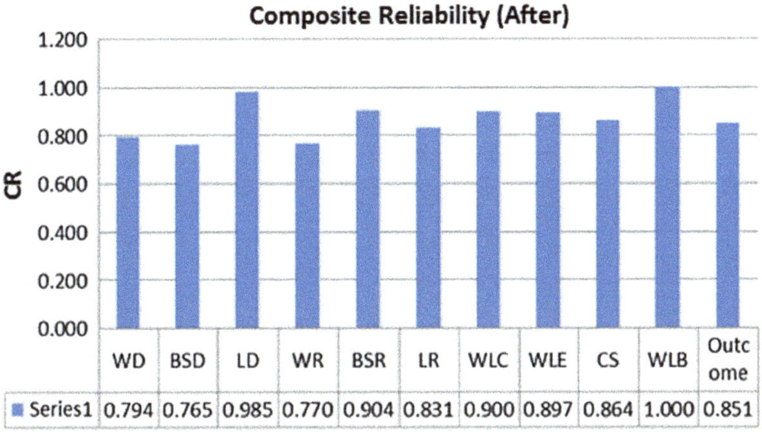

Fig. 6.6 Results of composite reliability for Singapore (after deletion)

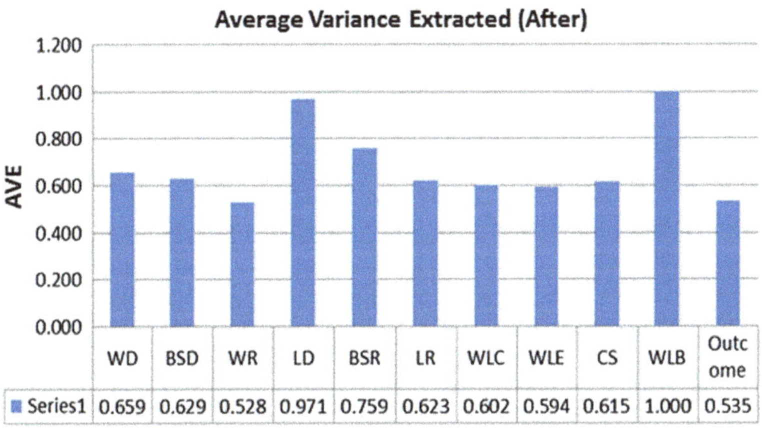

Fig. 6.7 Results of average variance extracted for Singapore (after deletion)

removed resulting in the square root AVE for the BSD construct (0.793) greater than the correlations of other constructs. With all the statistical criteria met, the validity and the reliability of the reflective measurement model were established for the Singapore results.

Table 6.23 Results of cross-loadings (after deletion)

Cross-Loadings											
	BSD	BSR	CS	LD	LR	OC	WD	WLB	WLC	WLE	WR
BoundarySD1	0.620	0.295	-0.365	-0.008	-0.012	-0.282	0.347	-0.310	0.339	-0.290	-0.174
BoundarySD4	0.934	-0.003	-0.040	0.171	0.109	-0.016	0.072	0.252	0.161	-0.043	-0.123
BoundarySR2	0.115	0.911	-0.101	0.345	-0.339	-0.119	-0.066	-0.134	0.036	-0.345	-0.345
BoundarySR3	0.094	0.797	-0.123	0.305	-0.355	-0.054	-0.038	-0.027	-0.088	-0.362	-0.282
BoundarySR4	0.061	0.900	-0.006	0.363	-0.355	-0.013	-0.114	-0.121	0.064	-0.249	-0.185
CS1	-0.141	-0.066	0.823	0.141	0.498	0.324	-0.115	0.546	-0.044	0.422	0.479
CS2	-0.170	-0.142	0.849	0.197	0.251	0.458	-0.171	0.403	-0.012	0.636	0.457
CS4	0.037	0.100	0.730	0.356	0.046	0.508	-0.078	0.431	0.131	0.447	0.294
CS5	-0.239	-0.183	0.721	0.078	0.127	0.368	-0.195	0.502	-0.068	0.491	0.276
LDemand2	0.131	0.380	0.256	0.989	0.183	-0.078	0.185	0.044	0.081	0.293	0.283
LDemand3	0.142	0.382	0.201	0.982	0.185	-0.088	0.227	-0.063	0.064	0.252	0.295
LResource1	0.005	-0.333	0.064	-0.029	0.729	0.163	0.106	0.290	0.051	0.463	0.213
LResource2	0.112	-0.258	0.289	0.183	0.857	0.291	0.364	0.316	0.142	0.602	0.455
LResource3	0.071	-0.366	0.365	0.238	0.776	0.143	0.095	0.100	0.024	0.668	0.520
OutcomeLife1	-0.163	0.024	0.349	-0.255	-0.121	0.663	-0.306	0.335	-0.040	0.231	0.140
OutcomeLife2	-0.067	0.197	0.299	-0.012	-0.010	0.657	0.075	0.325	0.094	0.075	0.135
OutcomeLife3	-0.012	-0.100	0.613	-0.055	0.371	0.854	-0.144	0.626	0.042	0.489	0.321
OutcomeWork1	-0.179	-0.248	0.567	-0.011	0.309	0.708	-0.008	0.298	-0.080	0.604	0.429
OutcomeWork2	-0.099	-0.176	0.558	0.061	0.268	0.756	0.119	0.226	0.049	0.484	0.547
WDemand2	0.497	-0.197	-0.159	0.098	0.225	-0.143	0.759	-0.026	0.557	-0.103	0.028
WDemand3	0.315	0.040	-0.134	0.223	0.175	-0.043	0.862	0.222	0.070	-0.047	0.057
WLC1	0.500	-0.102	-0.051	-0.115	0.198	0.117	0.499	0.304	0.761	0.020	0.125
WLC2	0.586	-0.027	0.077	-0.058	-0.080	0.124	0.438	0.174	0.738	-0.052	-0.141
WLC3	0.396	-0.077	-0.009	0.133	0.113	0.007	0.085	-0.172	0.855	-0.081	-0.012
WLC4	0.410	0.143	0.005	0.138	-0.050	-0.176	0.505	0.277	0.758	-0.169	-0.099
WLC5	0.282	0.214	0.253	0.366	0.193	0.182	0.473	-0.098	0.673	0.120	0.245
WLC6	0.600	-0.033	-0.193	-0.030	0.046	-0.111	0.076	0.250	0.855	-0.244	-0.209
WLE1	-0.119	-0.288	0.488	0.289	0.433	0.332	-0.105	0.375	-0.072	0.716	0.639
WLE2	-0.133	-0.307	0.494	0.237	0.435	0.351	-0.094	0.257	-0.216	0.780	0.526
WLE3	-0.220	-0.306	0.431	0.174	0.440	0.299	-0.279	0.438	-0.097	0.726	0.445
WLE4	0.023	-0.363	0.395	0.166	0.436	0.363	0.126	0.402	0.040	0.738	0.491
WLE5	-0.007	-0.195	0.245	0.241	0.316	0.525	0.117	0.402	0.094	0.821	0.555
WLE6	-0.235	-0.299	0.565	0.177	0.530	0.484	-0.215	0.455	-0.264	0.834	0.455
WResource1	-0.084	-0.321	0.537	0.174	0.475	0.294	-0.044	0.216	-0.043	0.600	0.673
WResource2	-0.210	-0.237	0.231	0.109	0.311	0.448	0.076	0.238	0.035	0.400	0.728
WResource3	-0.083	-0.091	0.198	0.367	0.292	0.229	0.128	0.230	-0.069	0.395	0.775
WorkLifeBalance	0.091	-0.104	0.604	-0.003	0.285	0.554	-0.010	1.000	0.151	0.473	0.315

Table 6.24 Results of the Fornell-Larker criterion (after deletion)

Discriminant validity (Fornell-Larcker criterion)											
	BSD	BSR	CS	LD	LR	OC	WD	WLB	WLC	WLE	WR
BSD	0.793										
BSR	0.108	0.871									
CS	−0.169	−0.098	0.784								
LD	0.138	0.386	0.235	0.985							
LR	0.085	−0.404	0.326	0.187	0.789						
Outcome	−0.118	−0.077	0.550	−0.084	0.252	0.731					
WD	0.722	−0.079	−0.177	0.206	0.242	−0.106	0.812				
WLB	0.091	−0.104	0.604	−0.003	0.285	0.554	−0.010	1.000			
WLC	0.740	0.004	−0.007	0.075	0.091	0.023	0.790	0.151	0.776		
WLE	−0.143	−0.377	0.667	0.279	0.746	0.515	−0.087	0.473	−0.101	0.771	
WR	−0.167	−0.323	0.494	0.282	0.292	0.446	0.055	0.215	−0.039	0.670	0.726

6.12 Assessment of Reflective Measurement Model on South Korea Data

6.12.1 Assessing the Statistical Criterion of Reflective Measurement Model Before Deletion

Figure 6.8 presents the outer loadings and the path coefficients of the reflective model before deletion. The first criterion evaluated was the indicator reliability. Based on the results of the Outer Loadings shown in Table 6.25, it could be seen that there were a few indicators with outer loadings less than 0.708. Values highlighted in red are loadings >0.40 and would be removed from the construct. These indicators are: BoundarySD3 (0.276), WResource3 (0.259), BoundarySR2 (0.083), CS5 (0.034) and OutcomeWork3 (−0.013). Items highlighted in green were further examined to determine if the deletion would lead to an increase in other statistical criterion. The following indicators have outer loadings ranging from 0.40 to 0.70: WDemand1 (0.632), BoundarySD4 (0.641), LDemand3 (0.628), WResource1 (0.561), BoundarySR4 (0.567), LResource3 (0.648), WLC1 (0.616), WLC5 (0.507), WLC6 (0.585), CS1 (0.650) and CS5 (0.627). Composite Reliability was used to test the internal consistency reliability of the model. A rule of thumb for establishing internal consistency reliability is that the Composite Reliability values of each construct must be 0.70 and above. Figure 6.9 shows the Composite Reliability of the test results. In order to establish internal consistency reliability, each construct must have a Composite Reliability of above 0.70. However, it could be seen that all the constructs have Composite Reliability of more than 0.7 except WR that has a Composite Reliability of 0.62. The next criterion evaluated was the convergent validity and the method of evaluation was the AVE. To establish convergent validity, the AVE value must be above 0.50. Figure 6.10 showed that BSD (0.433), BSR (0.477), CS (0.392), WLC (0.425) and WR (0.401) have AVE values less than 0.50. This means the average outer loadings of the construct were less than 0.708 and these would be adjusted after the removal of some indicators.

Discriminant validity was the final criterion evaluated. There were two approaches to assess the discriminate validity: Cross-Loadings and the Fornell-Larcker criterion. Table 6.26 presents the results of Cross-Loadings for South Korea. It could be seen that all the indicator's outer loadings from the same construct have values greater than the cross-loadings of other indicators. The indicators (outer loadings <0.40) highlighted in grey were not assessed as these were assumed to be removed prior to the evaluation. The second approach used to assess the discriminant validity was by using the Fornell-Larcker criterion. Table 6.27 presents the results of the Fornell-Larcker criterion of the reflective measurement model. It could be seen that most of the square root of each construct's AVE were greater than the highest correlation of other constructs except for the Outcome construct. The square root of the Outcome construct's AVE was 0.727 and the highest correlation value was the WLB construct (0.758). Thus, the

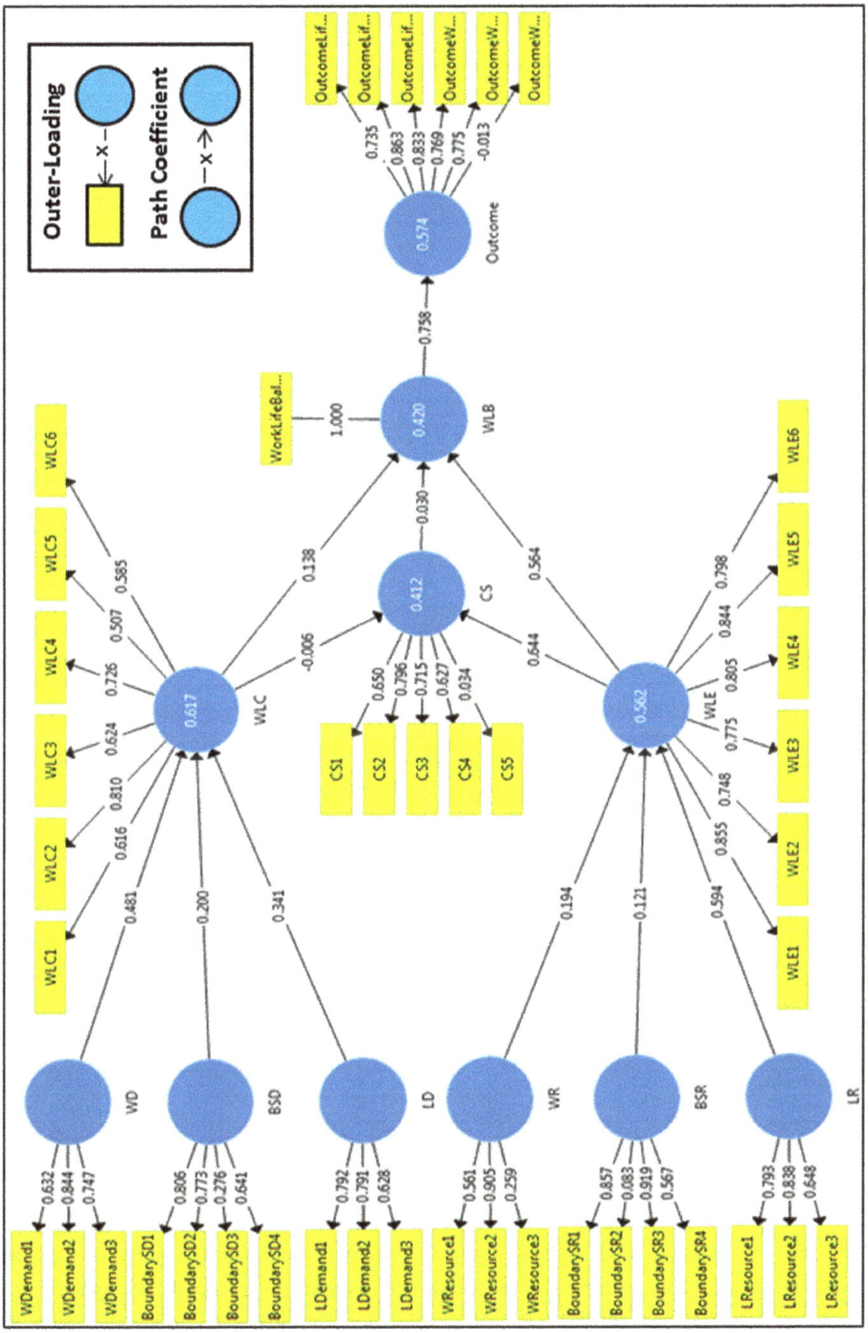

Fig. 6.8 Outer loadings and path coefficients of the reflective measurement model for South Korea result (before deletion)

Table 6.25 Results of the outer loadings before deletion (South Korea)

Outer Loadings											
	WD	BSD	LD	WR	BSR	LR	WLC	WLE	CS	WLB	Outcome
WDemand1	0.632										
WDemand2	0.844										
WDemand3	0.747										
BoundarySD1		0.806									
BoundarySD2		0.773									
BoundarySD3		0.276									
BoundarySD4		0.641									
LDemand1			0.792								
LDemand2			0.791								
LDemand3			0.628								
WResource1				0.561							
WResource2				0.905							
WResource3				0.259							
BoundarySR1					0.857						
BoundarySR2					0.083						
BoundarySR3					0.919						
BoundarySR4					0.567						
LResource1						0.793					
LResource2						0.838					
LResource3						0.648					
WLC1							0.616				
WLC2							0.810				
WLC3							0.862				
WLC4							0.726				
WLC5							0.507				
WLC6							0.585				
WLE1								0.855			
WLE2								0.748			
WLE3								0.775			
WLE4								0.805			
WLE5								0.844			
WLE6								0.798			
CS1									0.650		
CS2									0.796		
CS3									0.715		
CS4									0.627		
CS5									0.034		
WorkLifeBalance										1.000	
OutcomeWork1											0.769
OutcomeWork2											0.775
OutcomeWork3											-0.013
OutcomeLife1											0.735
OutcomeLife2											0.863
OutcomeLife3											0.833

measurement model was unable to establish the discriminant validity and would be reassessed after the removal of the Outcome construct's indicator (lowest outer-loading).

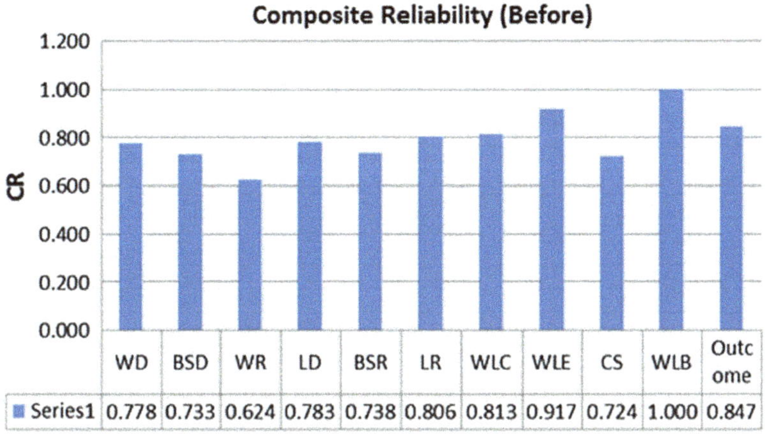

Fig. 6.9 Results of composite reliability for South Korea (before deletion)

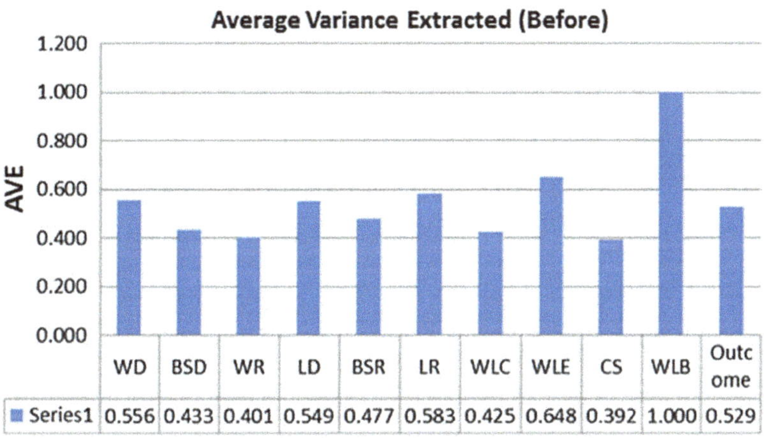

Fig. 6.10 Results of average variance extracted for South Korea (before deletion)

6.12.2 Assessing the Statistical Criterion of Reflective Measurement Model After Deletion

After assessing the statistical criterion, some of the observed variables were removed in order to meet the requirements stipulated by the criterion. Table 6.28 shows the outer loading results after deletion. It could be seen that BoundarySD3, WResource3, BoundarySR2, CS4, CS5, WLC1, WLC5 and OutcomeWork3 were removed from their constructs. Figure 6.11 shows the outer loadings model and the path coefficients of the overall model after deletion. Six observed variables in total were removed in order to preserve the validity and reliability of the model. There

Table 6.26 Results of cross-loadings (before deletion)

Cross-Loadings											
	BSD	BSR	CS	LD	LR	OC	WD	WLB	WLC	WLE	WR
BoundarySD1	0.806	0.041	0.229	0.139	0.309	0.164	0.154	0.266	0.333	0.196	0.141
BoundarySD2	0.773	-0.081	-0.087	0.052	0.094	0.119	0.110	0.257	0.221	-0.038	0.089
BoundarySD3	0.276	-0.107	-0.271	0.162	-0.207	-0.233	-0.045	-0.055	0.133	-0.120	-0.134
BoundarySD4	0.641	0.058	-0.056	0.485	-0.042	0.059	0.132	0.119	0.280	0.109	0.372
BoundarySR1	-0.112	0.857	0.392	0.237	0.367	0.197	-0.038	0.140	0.069	0.415	0.110
BoundarySR2	0.337	0.083	-0.006	0.096	0.192	-0.206	-0.042	-0.125	-0.063	-0.038	-0.088
BoundarySR3	-0.069	0.919	0.320	0.134	0.459	0.271	-0.001	0.331	0.052	0.350	0.094
BoundarySR4	0.244	0.567	0.181	0.038	0.423	0.269	-0.112	0.443	0.040	0.329	0.318
CS1	-0.132	0.370	0.650	0.080	0.398	0.240	0.029	0.283	-0.008	0.333	0.191
CS2	-0.134	0.404	0.796	0.263	0.435	0.438	0.206	0.352	0.188	0.505	0.074
CS3	0.157	0.027	0.715	0.407	0.342	0.394	0.350	0.240	0.374	0.527	0.339
CS4	0.123	0.310	0.627	0.040	0.406	0.340	-0.057	0.313	0.050	0.407	0.117
CS5	0.060	0.074	0.034	-0.015	-0.023	-0.131	-0.162	-0.097	-0.001	0.061	-0.057
LDemand1	0.240	0.315	0.477	0.791	0.432	0.487	0.441	0.382	0.432	0.507	0.176
LDemand2	0.448	0.236	0.214	0.791	0.369	0.302	0.196	0.422	0.466	0.394	0.259
LDemand3	0.045	-0.131	-0.003	0.628	0.001	0.087	0.397	0.011	0.489	0.030	0.258
LResource1	0.194	0.381	0.482	0.374	0.793	0.345	-0.036	0.418	0.077	0.663	0.256
LResource2	-0.027	0.458	0.396	0.167	0.838	0.667	0.238	0.680	0.232	0.582	0.210
LResource3	0.144	0.351	0.434	0.273	0.648	0.647	0.161	0.597	0.140	0.286	0.297
OutcomeLife1	-0.090	0.133	0.391	0.270	0.405	0.735	0.293	0.599	0.306	0.452	0.134
OutcomeLife2	0.192	0.210	0.505	0.206	0.626	0.863	0.183	0.660	0.287	0.484	0.326
OutcomeLife3	0.100	0.350	0.543	0.474	0.421	0.833	0.316	0.634	0.295	0.577	0.394
OutcomeWork1	0.090	0.081	0.179	0.398	0.379	0.769	0.434	0.579	0.408	0.406	0.314
OutcomeWork2	0.076	0.510	0.401	0.181	0.496	0.775	0.121	0.526	0.190	0.494	0.383
OutcomeWork3	0.288	0.043	0.107	0.006	0.099	-0.013	0.013	0.052	0.250	0.263	-0.087
WDemand1	0.067	-0.217	0.129	0.172	0.080	0.101	0.632	0.161	0.377	0.182	0.149
WDemand2	0.333	-0.020	0.030	0.352	0.118	0.332	0.844	0.206	0.164	0.085	0.212
WDemand3	-0.126	0.063	0.359	0.514	0.097	0.286	0.747	0.021	0.447	0.164	0.100
WLC1	0.403	0.154	-0.127	0.259	0.088	0.304	0.271	0.352	0.616	0.170	0.384
WLC2	0.337	0.178	0.274	0.440	0.208	0.289	0.340	0.363	0.810	0.342	0.284
WLC3	0.292	-0.096	0.161	0.241	0.093	0.190	0.244	0.274	0.624	0.389	0.137
WLC4	0.201	0.014	0.200	0.420	0.122	0.318	0.593	0.145	0.726	0.157	0.144
WLC5	0.064	-0.086	0.008	0.533	-0.025	0.126	0.417	-0.010	0.507	0.038	0.378
WLC6	0.279	0.093	0.307	0.331	0.223	0.214	0.158	0.312	0.585	0.415	0.067
WLE1	0.164	0.418	0.549	0.371	0.615	0.452	0.172	0.431	0.332	0.855	0.349
WLE2	0.000	0.409	0.373	0.259	0.434	0.462	0.121	0.584	0.313	0.748	0.384
WLE3	0.070	0.380	0.507	0.436	0.540	0.600	0.195	0.547	0.472	0.775	0.439
WLE4	0.188	0.378	0.492	0.370	0.492	0.418	0.089	0.451	0.225	0.805	0.304
WLE5	0.162	0.414	0.526	0.272	0.586	0.530	0.161	0.553	0.241	0.844	0.279
WLE6	-0.040	0.258	0.401	0.263	0.541	0.493	0.111	0.506	0.216	0.798	0.213
WResource1	0.367	0.128	0.274	0.276	0.054	0.185	0.108	0.217	0.238	0.201	0.561
WResource2	0.042	0.146	0.177	0.296	0.311	0.535	0.234	0.303	0.289	0.387	0.905
WResource3	0.262	0.261	-0.001	-0.179	0.245	0.144	-0.102	0.175	0.034	0.089	0.259
WorkLifeBalance	0.262	0.377	0.424	0.361	0.302	0.224	0.181	1.000	0.355	0.634	0.363

were two reasons for the deletion: poor outer loadings and poor convergent validity. In poor outer loadings, loadings with less than 0.40 were automatically removed (Hair et al., 2017). These indicators include BoundarySD3, WResource3, BSR2, CS5 and OutcomeWork3. Next, in order to establish convergent validity, the AVE

Table 6.27 Results of the Fornell-Larker criterion (before deletion)

Discriminant Validity (Fornell-Larcker Criterion)											
	BSD	BSR	CS	LD	LR	OC	WD	WLB	WLC	WLE	WR
BSD	0.658										
BSR	-0.004	0.690									
CS	0.008	0.389	0.626								
LD	0.329	0.179	0.299	0.741							
LR	0.129	0.517	0.562	0.353	0.764						
Outcome	0.101	0.317	0.516	0.387	0.669	0.727					
WD	0.163	-0.060	0.206	0.467	0.134	0.340	0.746				
WLB	0.262	0.377	0.424	0.361	0.704	0.758	0.181	1.000			
WLC	0.391	0.073	0.234	0.632	0.189	0.380	0.673	0.355	0.652		
WLE	0.109	0.470	0.642	0.410	0.718	0.613	0.177	0.634	0.372	0.805	
WR	0.233	0.217	0.254	0.317	0.315	0.529	0.212	0.363	0.335	0.407	0.633

of the construct must be 0.50 and above. After the deletion of outer loadings less than 0.40, it was left with the CS and WLC constructs that were unable to meet the AVE target of >0.50; hence the lowest outer loading from the construct was first removed. WLC5 and CS5 were initially removed; resulting in the AVE values of the WLC and CS constructs to be 0.487 and 0.49 respectively. These were lower than the 0.50 target for convergent validity. Hence, the next lowest loading WLC1 and CS4 were removed which resulted in the AVE of the WLC and CS constructs to be 0.530 and 0.582 respectively. With all the AVE of the constructs above 0.50, the convergent validity of the measurement model was established. Composite Reliability was already established prior to the deletion of the outer loadings. In summary, Figs. 6.12 and 6.13 show the AVE and Composite Reliability of the constructs. These showed that all the constructs have met the criterion stated by Hair et al. (2017). The internal consistency reliability and convergent validity were established for the measurement model.

Discriminant validity was the final criterion evaluated using Cross-Loadings and the Fornell-Larcker criterion. Table 6.26 shows that there were no cross-loadings greater than the loading of the construct's own outer loading. This means the discriminant validity of the indicators has already been established (before deletion). In Table 6.29, after the removal of the eight indicators, there were no major changes to the cross-loadings and all the loadings from their own construct were still greater than the cross-loadings. The second approach to assess the discriminant validity used the Fornell-Larcker criterion. Table 6.27 shows that before deletion,

Table 6.28 Results for the outer loadings after deletion (South Korea)

Outer loadings											
	WD	BSD	LD	WR	BSR	LR	WLC	WLE	CS	WLB	Outcome
WDemand1	0.622										
WDemand2	0.844										
WDemand3	0.755										
BoundarySD1		0.888									
BoundarySD2		0.845									
BoundarySD4		0.511									
LDemand1			0.854								
LDemand2			0.833								
LDemand3			0.512								
WResource1				0.597							
WResource2				0.907							
BoundarySR1					0.860						
BoundarySR3					0.921						
BoundarySR4					0.564						
LResource1						0.791					
LResource2						0.839					
LResource3						0.649					
WLC2							0.802				
WLC3							0.707				
WLC4							0.723				
WLC6							0.672				
WLE1								0.854			
WLE2								0.751			

(continued)

Table 6.28 (continued)

Outer loadings											
	WD	BSD	LD	WR	BSR	LR	WLC	WLE	CS	WLB	Outcome
WLE3								0.781			
WLE4								0.803			
WLE5								0.842			
WLE6								0.794			
CS1									0.679		
CS2									0.866		
CS3									0.731		
WorkLifeBalance										1.000	
OutcomeWork1											0.770
OutcomeWork2											0.775
OutcomeLife1											0.737
OutcomeLife2											0.861
OutcomeLife3											0.833

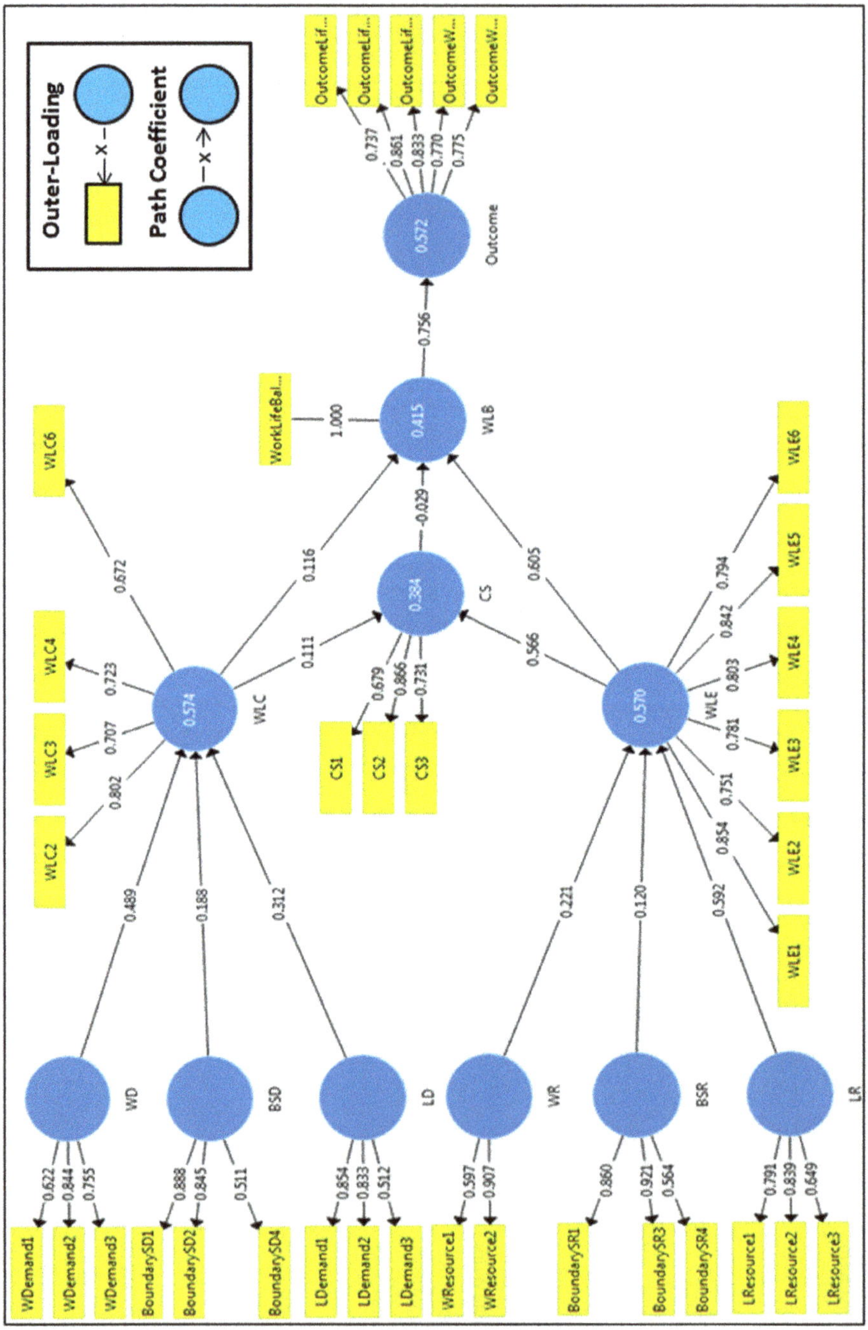

Fig. 6.11 Results of outer loadings and path coefficients of the reflective measurement model for South Korea (after deletion)

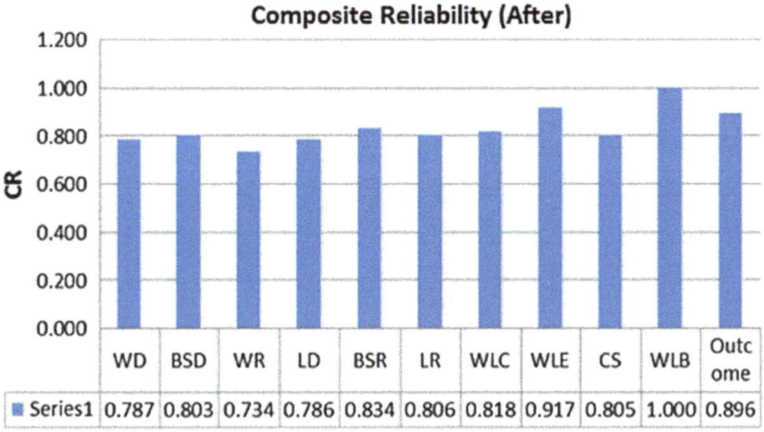

Fig. 6.12 Results of composite reliability for South Korea (after deletion)

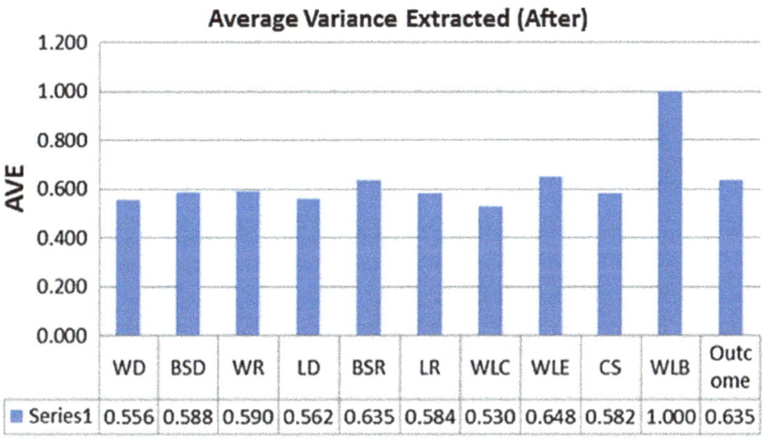

Fig. 6.13 Results of average variance extracted for South Korea (after deletion)

the square root AVE of the Outcome construct (0.727) was lower than the correlation of WLB construct (0.758). In Table 6.30, after the removal of the OutcomeWork3 indicator, the squared AVE of Outcome construct (0.797) was greater than the correlation of the WLB construct (0.756) resulting in the model achieving discriminant validity. With all the statistical criteria met, the validity and the reliability of the reflective measurement model were established for South Korea.

Table 6.29 Results of cross-loadings (after deletion)

Cross-Loadings											
	BSD	BSR	CS	LD	LR	OC	WD	WLB	WLC	WLE	WR
BoundarySD1	0.888	0.059	0.144	0.167	0.309	0.156	0.151	0.266	0.364	0.196	0.082
BoundarySD2	0.845	-0.070	-0.147	0.083	0.094	0.116	0.107	0.257	0.224	-0.038	0.004
BoundarySD4	0.511	0.062	-0.028	0.475	-0.042	0.057	0.132	0.119	0.188	0.109	0.398
BoundarySR1	-0.104	0.860	0.372	0.255	0.367	0.195	-0.033	0.140	0.077	0.415	0.093
BoundarySR3	-0.070	0.921	0.285	0.172	0.460	0.270	-0.002	0.331	0.060	0.351	0.065
BoundarySR4	0.279	0.564	0.102	0.091	0.423	0.268	-0.113	0.443	0.030	0.331	0.255
CS1	-0.040	0.368	0.679	0.095	0.399	0.244	0.032	0.283	0.047	0.336	0.155
CS2	-0.092	0.405	0.866	0.286	0.435	0.437	0.211	0.352	0.243	0.508	0.088
CS3	0.171	0.025	0.731	0.433	0.341	0.392	0.350	0.240	0.437	0.525	0.382
LDemand1	0.206	0.314	0.507	0.854	0.432	0.487	0.447	0.382	0.493	0.507	0.215
LDemand2	0.385	0.239	0.192	0.833	0.368	0.301	0.197	0.422	0.466	0.395	0.259
LDemand3	-0.024	-0.126	0.099	0.512	0.002	0.088	0.399	0.011	0.342	0.032	0.315
LResource1	0.207	0.396	0.393	0.413	0.791	0.341	-0.037	0.418	0.132	0.660	0.245
LResource2	0.084	0.456	0.383	0.187	0.839	0.664	0.238	0.680	0.231	0.584	0.157
LResource3	0.216	0.348	0.443	0.305	0.649	0.651	0.163	0.597	0.143	0.287	0.279
OutcomeLife1	-0.013	0.118	0.392	0.293	0.406	0.737	0.296	0.599	0.311	0.453	0.158
OutcomeLife2	0.285	0.200	0.436	0.260	0.626	0.861	0.184	0.660	0.299	0.484	0.288
OutcomeLife3	0.112	0.351	0.527	0.456	0.421	0.833	0.319	0.634	0.336	0.577	0.380
OutcomeWork1	0.090	0.069	0.197	0.378	0.380	0.770	0.432	0.579	0.311	0.408	0.413
OutcomeWork2	0.121	0.504	0.352	0.201	0.497	0.775	0.123	0.526	0.152	0.497	0.447
WDemand1	0.080	-0.219	0.196	0.141	0.080	0.102	0.622	0.161	0.358	0.182	0.184
WDemand2	0.318	-0.017	0.056	0.349	0.119	0.328	0.844	0.206	0.624	0.085	0.211
WDemand3	-0.090	0.057	0.444	0.505	0.098	0.289	0.755	0.021	0.455	0.167	0.134
WLC2	0.349	0.178	0.300	0.425	0.208	0.286	0.339	0.363	0.802	0.345	0.278
WLC3	0.272	-0.102	0.138	0.255	0.092	0.181	0.239	0.274	0.707	0.389	0.136
WLC4	0.193	0.015	0.253	0.438	0.123	0.316	0.537	0.145	0.723	0.158	0.162
WLC6	0.260	0.089	0.284	0.363	0.222	0.205	0.155	0.312	0.672	0.416	0.075
WLE1	0.160	0.319	0.531	0.408	0.614	0.445	0.172	0.431	0.391	0.854	0.358
WLE2	0.042	0.394	0.381	0.263	0.434	0.459	0.120	0.584	0.265	0.751	0.376
WLE3	0.093	0.378	0.566	0.460	0.540	0.596	0.196	0.547	0.485	0.781	0.403
WLE4	0.180	0.390	0.439	0.402	0.492	0.412	0.088	0.451	0.273	0.803	0.302
WLE5	0.008	0.405	0.561	0.325	0.586	0.527	0.160	0.553	0.311	0.842	0.291
WLE6	0.197	0.250	0.316	0.310	0.541	0.486	0.111	0.506	0.236	0.794	0.218
WResource1	0.338	0.124	0.272	0.261	0.054	0.186	0.106	0.217	0.187	0.204	0.597
WResource2	0.034	0.140	0.196	0.278	0.311	0.537	0.233	0.303	0.188	0.388	0.907
WorkLifeBalance	0.291	0.370	0.381	0.395	0.304	0.273	0.179	1.000	0.355	0.635	0.342

6.13 Assessment of Structural Model on Singapore Data

After the validity and reliability of the measurement model were established, the structural model was addressed. The structural model was used to determine the model's capabilities in predicting the relationships between the constructs as well as to test the hypotheses formulated for the research. The six steps used in assessing the structural model are shown in Fig. 6.14.

The first step in assessing the structural model was the examination for collinearity. The reason for measuring collinearity was that the measurement of path coefficients in the structural model was based on the Ordinary Least Squares regressions of each latent variable (endogenous) on its own predecessor constructs

Table 6.30 Results of the Fornell-Larker criterion (after deletion)

Discriminant validity (Fornell-Larcker criterion)											
	BSD	BSR	CS	LD	LR	OC	WD	WLB	WLC	WLE	WR
BSD	0.767										
BSR	0.028	0.797									
CS	0.023	0.333	0.763								
LD	0.278	0.227	0.379	0.749							
LR	0.206	0.523	0.509	0.391	0.764						
Outcome	0.153	0.305	0.483	0.417	0.667	0.797					
WD	0.171	−0.058	0.281	0.456	0.135	0.341	0.746				
WLB	0.291	0.370	0.381	0.395	0.704	0.756	0.179	1.000			
WLC	0.358	0.073	0.344	0.587	0.219	0.358	0.663	0.355	0.728		
WLE	0.136	0.467	0.611	0.452	0.716	0.608	0.178	0.635	0.411	0.805	
WR	0.173	0.168	0.277	0.340	0.278	0.520	0.237	0.342	0.234	0.406	0.768

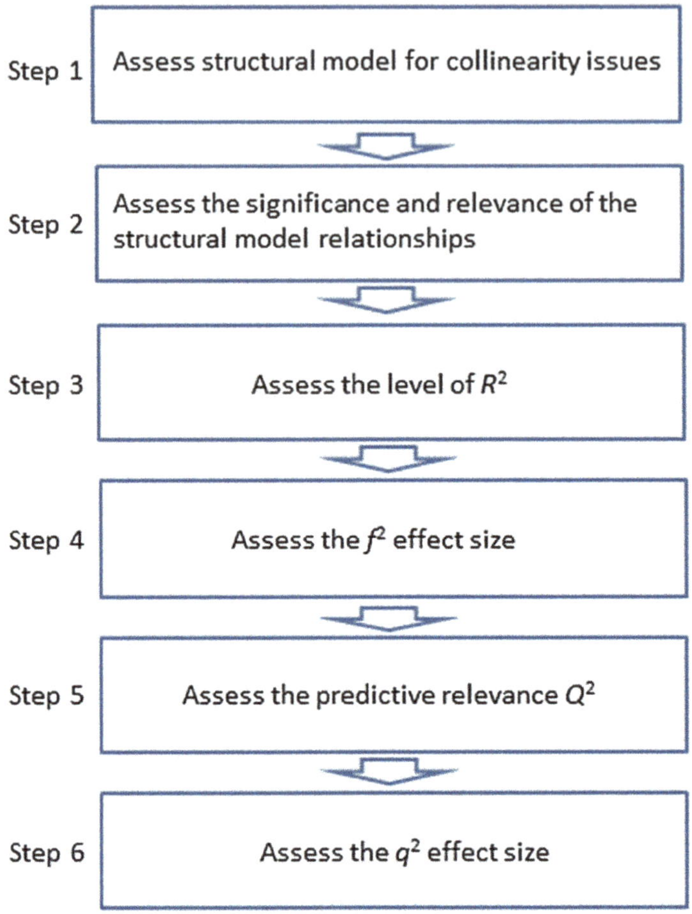

Fig. 6.14 Structural model assessment procedure

(Hair et al., 2017). If there were severe levels detected in the collinearity among the predictor constructs, removal of the construct might be inevitable. To assess the collinearity issues, VIF inner values of all sets of predictor constructs must be examined. The rule of thumb is the VIF values must be below 5 in order for the collinearity not to be a critical issue in the construct. Table 6.31 presents the VIF values for all the different combinations of endogenous (represented by the rows) and exogenous constructs (represented by the columns). In this model, five sets of path were assessed for collinearity: (1) WD, BSD and LD as predictors of WLC; (2) WR, BSR and LR as predictors of WLE; (3) WLC, WLE as predictors of CS; (4) WLC, WLE and CS as predictors of WLB; and (5) WLB as predictor of Outcome. As shown in Table 6.31, all VIF values were well below the threshold of 5. Thus, collinearity among the predictor constructs was not considered a critical issue in the structural model.

Table 6.31 VIF inner values in the structural model (Singapore)

Collinearity statistics (inner VIF)											
	BSD	BSR	CS	LD	LR	OC	WD	WLB	WLC	WLE	WR
BSD									2.088		
BSR										1.220	
CS								2.473			
LD									1.045		
LR										1.507	
OC											
WD									2.140		
WLB						1.000					
WLC			1.010					1.023			
WLE			1.010					2.498			
WR										1.409	

Next, the significance and relevancy of the structural model relationships were assessed. In this step, two important elements were assessed: Path Coefficient and *t*-statistics (or significance level). Figure 6.15 shows the *t*-statistics of the entire structural model. However, for assessment of the structural model, only the *t*-values between constructs (latent variables) would be required for evaluation. The estimates were obtained after running the PLS-SEM algorithms and the path coefficients represent the hypothesized relationships between the constructs. The path coefficient has a value ranging from −1 to 1 where a value close to 1 was considered statistically significant (Hair et al., 2017) and a value close to 0 was considered not significant. Table 6.32 presents the path coefficients between constructs. It could be seen that the highest path coefficient value was CS → WLB (0.561) which means that CS was an important construct to explain WLB. To next assess whether all the relationships between constructs were significant, the Bootstrapping interface was run. Bootstrapping is a nonparametric procedure where subsamples are randomly and automatically drawn from the data. This allows testing of the statistical significance of the path coefficients (Hair et al., 2017). When using the Bootstrapping interface, one has to note that the results of the *t*-statistics or other data would be different as it uses different subsamples for generating results. In this analysis, Hair et al. (2017) recommendations were adopted with subsamples = 5000, two-tailed testing and significance level of 0.05. Table 6.33 summarizes the structural model's Path Coefficients, *t*-values and *p* values.

At the 5% significance level, it could be seen that all the relationships in the structural model were significant except BSR → WLE ($p = 0.721$), LD → WLC ($p = 0.487$), WLC → CS ($p = 0.476$), WLC → WLB ($p = 0.310$) and WLE → WLB ($p = 0.812$). It is important to note that the *t*-values for the significance level of 5% must be greater than 1.96. The results suggested that for the Singaporeans to attain WLB, they would require Coping Strategies with Work–Life Enrichment as its statistically significant predecessor. Increasing WLE or reducing

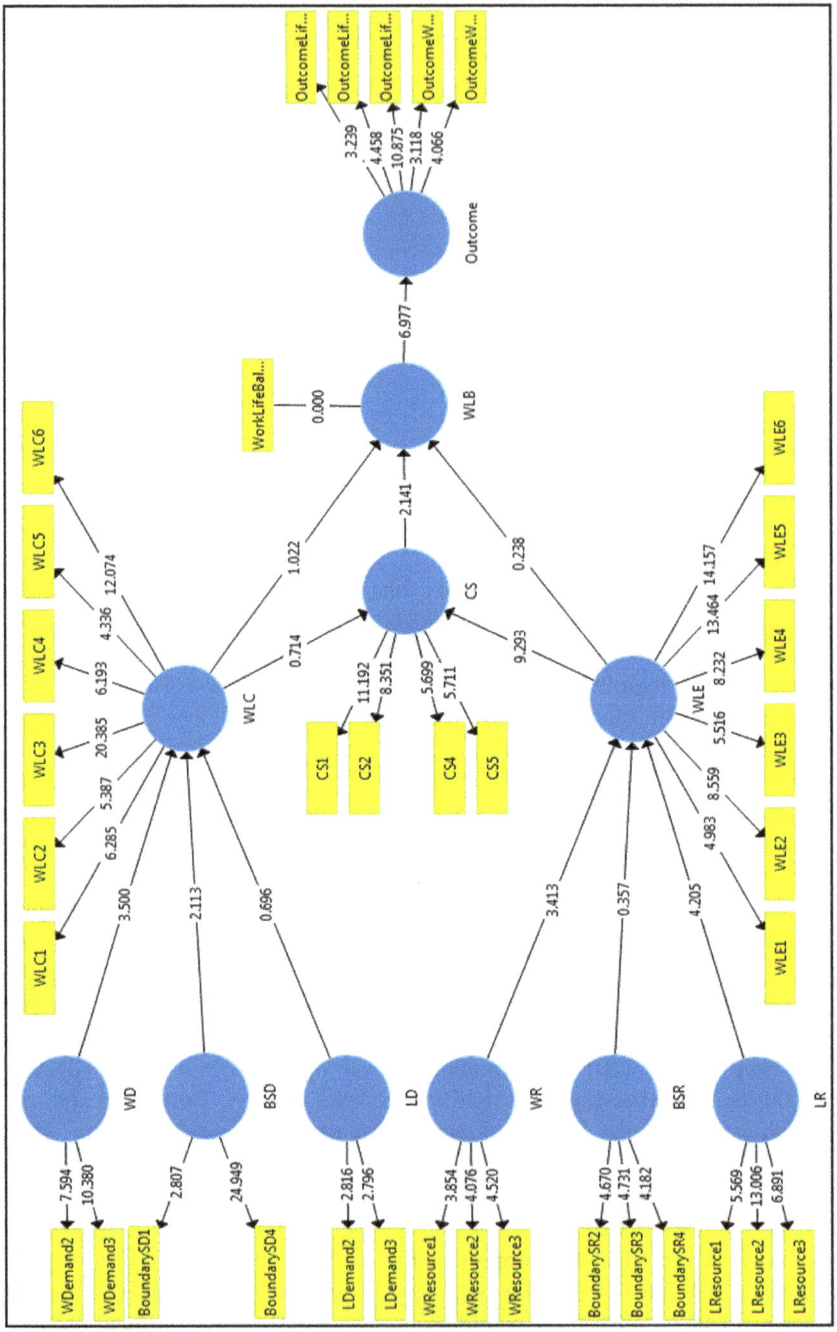

Fig. 6.15 Results of Bootstrapping *t*-values of the measurement and structural model relationships for Singapore

Table 6.32 Path coefficients of the structural model

Path coefficients											
	BSD	BSR	CS	LD	LR	OC	WD	WLB	WLC	WLE	WR
BSD									0.352		
BSR										−0.039	
CS								0.561			
LD									−0.088		
LR										0.527	
OC											
WD									0.555		
WLB						0.554					
WLC			0.072					0.161			
WLE			0.776					0.058			
WR										0.387	

Table 6.33 Summary of the path coefficients, *t*-values and *p* values (Singapore)

	Path coefficients	*t*-values	*p* values	Significance ($p < 0.05$)?
BSD → WLC	0.352	2.113	0.035	Yes
BSR → WLE	−0.039	0.357	0.721	No
CS → WLB	0.561	2.141	0.032	Yes
LD → WLC	−0.088	0.696	0.487	No
LR → WLE	0.527	3.205	0.000	Yes
WD → WLC	0.555	3.500	0.001	Yes
WLB → Outcome	0.554	6.977	0.000	Yes
WLC → CS	0.072	0.714	0.476	No
WLC → WLB	0.161	1.016	0.310	No
WLE → CS	0.776	9.293	0.000	Yes
WLE → WLB	0.006	0.238	0.812	No
WR → WLE	0.387	3.413	0.001	Yes

WLC were insignificant in explaining WLB. Work Demand and Boundary Spanning Demand were statistically significant in explaining Work–Life Conflict; Work Resources and Life Resources were statistically significant in explaining Work–Life Enrichment; and Work–Life Balance was statistically significant in explaining Outcome. A more detailed evaluation of the research hypotheses would be explained later.

The path coefficient was used to evaluate direct exogenous constructs on the endogenous latent variables but researchers might also look at the indirect effects via other constructs on the endogenous latent variables (Hair et al., 2017). The sum of the direct and indirect effects is known as the total effect. Total effect is useful in studies that aimed at exploring different impact of the construct on the criterion construct via one or more mediating variables (Hair et al., 2017). Table 6.34 shows

Table 6.34 Summary of test of significance results of the total effects (Singapore)

	Total Effects	*t*-values	*p* values	Significance ($p < 0.05$)?
BSD → CS	0.025	0.569	0.570	No
BSD → Outcome	0.039	0.738	0.461	No
BSD → WLB	0.071	0.823	0.411	No
BSD → WLC	0.352	2.113	0.035	Yes
BSR → CS	-0.031	0.350	0.726	No
BSR → Outcome	-0.011	0.293	0.769	No
BSR → WLB	-0.019	0.328	0.743	No
BSR → WLE	-0.039	0.357	0.721	No
CS → Outcome	0.311	1.887	0.059	No
CS → WLB	0.561	2.141	0.032	Yes
LD → CS	-0.006	0.341	0.733	No
LD → Outcome	-0.010	0.420	0.674	No
LD → WLB	-0.018	0.469	0.639	No
LD → WLC	-0.088	0.696	0.487	No
LR → CS	0.409	3.628	0.000	Yes
LR → Outcome	0.144	2.364	0.018	Yes
LR → WLB	0.260	3.009	0.003	Yes
LR → WLE	0.527	4.205	0.000	Yes
WD → CS	0.040	0.689	0.485	No
WD → Outcome	0.062	1.017	0.309	No
WD → WLB	0.112	1.122	0.262	No
WD → WLC	0.555	3.500	0.001	Yes
WLB → Outcome	0.554	6.977	0.000	Yes
WLC→ CS	0.072	0.714	0.476	No
WLC → Outcome	0.112	0.999	0.318	No
WLC → WLB	0.201	1.109	0.268	No
WLE → CS	0.776	9.293	0.000	Yes
WLE → Outcome	0.273	3.119	0.002	Yes
WLE → WLB	0.493	4.244	0.000	Yes
WR → CS	0.300	3.313	0.001	Yes
WR → Outcome	0.106	2.185	0.029	Yes
WR → WLB	0.191	2.531	0.011	Yes
WR → WLE	0.387	3.413	0.001	Yes

a summary of the total effect values as well as the significance testing of the total effects. Rows highlighted in grey were indirect constructs that have significant impact on other endogenous constructs. The results showed that LR and WR as an antecedent of WLE have a high total effect on CS, Outcome and WLB and it has been posited that these two indicators were the ones that allowed the Singaporeans to achieve their WLB. One of the more interesting paths was between WLE → WLB as the path coefficient was only 0.006 but the total effect was substantial (0.493) and has a p value of 0.00 (at 5% significance level). This means that the indirect effect between WLE → WLB was statistically significant and it could be concluded that CS fully mediates WLE to WLB relationship. Hair et al. (2017) referred this as indirect-only mediation (full mediation).

The next evaluation of the structural model used the R^2 value or the coefficient of determination. R^2 measures the model's predictive accuracy and is calculated by squaring the correlation between the endogenous construct's predicted values (Hair et al., 2017). It also represents the exogenous constructs combined effects on the endogenous construct, indicates the amount of variance in the endogenous constructs and is explained by all the associated exogenous constructs linking to it (Hair et al., 2017). R^2 has a value ranging from 0 to 1 and till now there are no acceptable R^2 values as the result depends on the model complexity and the research objective. However, Hair et al. (2017) mentioned that R^2 values of 0.75, 0.50 and 0.25 for the endogenous constructs could be interpreted respectively as substantial, moderate and weak. Table 6.35 shows the R^2 values of all the endogenous latent variables. R^2 value for WLC (0.693) and WLC (0.669) constructs could be considered to be close to substantial; CS (0.596) construct was considered moderate; Outcome (0.307) and WLB (0.391) were considered to be fairly weak.

Evaluating the effect size of f^2 was the next step in assessing the structural model. In addition to calculating the R^2 values of all the endogenous latent variables, the change in the R^2 value after the omission of one of its linked exogenous construct could be used to evaluate whether removing that construct would lead to a substantive impact on the endogenous latent variables (Hair et al., 2017). This measure is the f^2 effect size and Table 6.36 shows a summary of the results. The rule of thumb in interpreting the f^2 values is as follows: 0.02 = small effects; 0.15 = medium effects; 0.35 = large effects; less than 0.02 = no effect (Hair et al., 2017). Following the rule of thumb set out by Hair et al. (2017), the results from Table 6.36 were interpreted as follows: BSD has medium effect on WLC (0.19);

Table 6.35 Summary of R^2 on the structural model (Singapore)

	R^2
CS	0.596
Outcome	0.307
WLB	0.391
WLC	0.693
WLE	0.669

Table 6.36 Summary of f^2 effect size (Singapore)

f^2 Effect size											
	BSD	BSR	CS	LD	LR	OC	WD	WLB	WLC	WLE	WR
BSD									0.193		
BSR										0.004	
CS								0.209			
LD									0.024		
LR										0.556	
OC											
WD									0.469		
WLB						0.444					
WLC			0.013					0.042			
WLE			1.473					0.002			
WR										0.321	

BSR has no effect on WLE (0.00); CS has medium effect on WLB (0.209); LD has small effect on WLC (0.02); WD has large effect on WLC (0.47); WLB has large effect on Outcome (0.44); WLC has no effect on CS (0.01) and small effect on WLB (0.04); WLE has large effect on CS (1.47) and no effect on WLB (0.00); and WR has medium effect on WLE (0.32).

The fifth step in evaluating the structural model was to examine the predictive relevance of Q^2 value. It has been posited that a Q^2 value of more than 0 for a specific reflective endogenous latent variable represents the path model's predictive relevancy for a dependent construct (Hair et al., 2017). The Q^2 value was obtained using the Blindfolding interface in SmartPLS. Blindfolding is an iterative process that repeats itself until each of the data point has been removed and the model re-estimated (Hair et al., 2017). Table 6.37 presents a summary of the Construct Cross-validated Redundancy estimates. SSO represents the sum of the square

Table 6.37 Summary of the Q^2 values (Singapore)

Construct cross-validated redundancy			
	SSO	SSE	Q^2 (1-SSE/SSO)
BSD	66.00	66.00	
BSR	99.00	99.00	
CS	132.00	91.02	0.3105
LD	66.00	66.00	
LR	99.00	99.00	
OC	165.00	148.22	0.1017
WD	66.00	66.00	
WLB	33.00	24.76	0.2494
WLC	198.00	137.96	0.3032
WLE	198.00	131.13	0.3377
WR	99.00	99.00	

observations (SSO) and SSE represents the sum of squared prediction errors (SSE). Q^2 was formulated to be 1-SSE/SSO and it should be noted that only endogenous constructs have Q^2 values. It could be seen that all five endogenous constructs have Q^2 value above zero (CS = 0.311, Outcome = 0.102, WLB = 0.249, WLC = 0.302 and WLE = 0.338). The results clearly supported the model's predictive relevancy regarding the endogenous latent variables.

The final step in evaluating the structural model was to assess the q^2 effect sizes. The use of q^2 was similar to the f^2 effect size approach for assessing R^2 where it assessed the change in the R^2 value after the omission of one of its linked exogenous construct. The effect size of q^2 also used this approach where it assessed an exogenous latent variable's contribution to an endogenous construct's Q^2 value. The formula for calculating q^2 is as follows:

$$q^2 = Q^2_{\text{included}} - Q^2_{\text{excluded}}/1 - Q^2_{\text{included}}$$

Q^2_{included} is the value shown in Table 6.38 based on the Blindfolding estimation. Q^2_{excluded} value was obtained by manually deleting the corresponding predecessor of the endogenous latent variable. For example, the WLC construct has an original Q^2 of 0.338 (Q^2_{included}) and the WD construct was a predecessor of WLC. When WD was deleted from the path model and re-estimated, the Q^2 of WLC construct was reduced to 0.303 (Q^2_{excluded}). By applying the formula: (0.303 − 0.223)/(1 − 0.303), the q^2 effect size of WD on WLC was 0.115. Hair et al. (2017) mentioned that q^2 value of 0.02, 0.15 or 0.35 could be interpreted that an exogenous construct has a small, medium or large predictive relevance respectively on its linked endogenous latent variable. This also means the q^2 effect size for the relationship between WD → WLC could be considered as close to medium predictive relevance.

Table 6.38 Summary of the Q^2_{excluded} and q^2 effect sizes (Singapore)

	$Q^2_{excluded}$	$Q^2_{included}$	q^2	Predictive relevance
BSD → WLC	0.292	0.3032	0.016	Small
BSR → WLE	0.339	0.3377	−0.002	Small
CS → WLB	0.183	0.2494	0.088	Small
LD → WLC	0.309	0.3032	−0.008	Small
LR → WLE	0.249	0.3377	0.134	~Medium*
WD → WLC	0.223	0.3032	0.115	Small
WLC → CS	0.317	0.3105	−0.010	Small
WLC → WLB	0.274	0.2494	−0.033	Small
WLE → CS	−0.016	0.3105	0.474	Large
WLE → WLB	0.275	0.2494	−0.034	Small
WR → WLE	0.387	0.3377	0.086	Small

*~ denotes close to medium or large

Table 6.39 Summary of the results for q^2 effect sizes (Singapore)

q^2 Effect sizes											
	BSD	BSR	CS	LD	LR	OC	WD	WLB	WLC	WLE	WR
BSD									0.016		
BSR										−0.002	
CS								0.088			
LD									−0.008		
LR										0.134	
OC											
WD									0.115		
WLB											
WLC			−0.010					−0.033			
WLE			0.474					−0.034			
WR										0.086	

Table 6.39 summarizes the results of the q^2 effect sizes on all the relationships in the structural model.

6.14 Assessment of Structural Model on South Korea Data

The first step in assessing South Korea's structural model was to examine for collinearity. To assess the collinearity issues, VIF inner values of all sets of predictor constructs must be examined. The rule of thumb was that the VIF values

Table 6.40 VIF Inner values in the structural model (South Korea)

Collinearity statistics (inner VIF)											
	BSD	BSR	CS	LD	LR	OC	WD	WLB	WLC	WLE	WR
BSD									1.086		
BSR										1.379	
CS								1.624			
LD									1.331		
LR										1.452	
OC											
WD									1.265		
WLB						1.000					
WLC			1.204					1.224			
WLE			1.204					1.724			
WR										1.085	

must be below 5 in order for the collinearity not to be a critical issue in the structural model. Table 6.40 shows the VIF inner values for all the different combinations of endogenous and exogenous constructs. All the VIF inner values were clearly below the threshold of 5. Therefore, collinearity among the predictor constructs was not considered a critical issue in the structural model.

Next, the significance and relevance of the structural model relationships were assessed. Two important approaches needed to be assessed: Path Coefficient and *t*-statistics. Path coefficient has a value ranging from −1 to 1 where a value close to 1 represents strong positive relationship and is usually statistically significant (Hair et al., 2017). Table 6.41 shows the path coefficients between the endogenous and exogenous constructs. The highest path coefficient value was WLE → WLB (0.605) which suggests that WLE was an important indicator in contributing to South Koreans' achievement of WLB. The lowest path coefficient was CS → WLB (−0.029) which indicates that CS contributed negligibly (or close to zero) in helping South Koreans achieved WLB. To test for the significance of the model, Bootstrapping interface was run. At the 5% significance level, it could be seen that all the relationships in the structural model were significant except BSD → WLC (0.152), BSR → WLE (0.238), CS → WLB (0.895), WLC → CS (0.536) and WLC → WLB (0.502). It was noted that *t*-values for significance level of 5% must be >1.96. The results suggested that for South Koreans to attain WLB, the organizations or parties concerned should concentrate on improving the WLE of individuals. In this context, Life Demands (LD) and Work Demands (WD) were statistically significant in explaining WLC. Life Resources (LR) and Work Resources (WR) were statistically significant in explaining WLE. WLE was statistically significant in explaining CS and WLE and finally WLB was statistically significant in explaining Outcome. Figure 6.16 shows the *t*-values for the measurement and structural model relationships. It was noted that only the *t*-values of the structural model relationship would be required for evaluation. Table 6.42

Table 6.41 Path coefficients of the structural model (South Korea)

Path coefficients											
	BSD	BSR	CS	LD	LR	OC	WD	WLB	WLC	WLE	WR
BSD									0.188		
BSR										0.120	
CS								−0.029			
LD									0.312		
LR										0.592	
OC											
WD									0.489		
WLB						0.756					
WLC			0.111					0.116			
WLE			0.566					0.605			
WR										0.221	

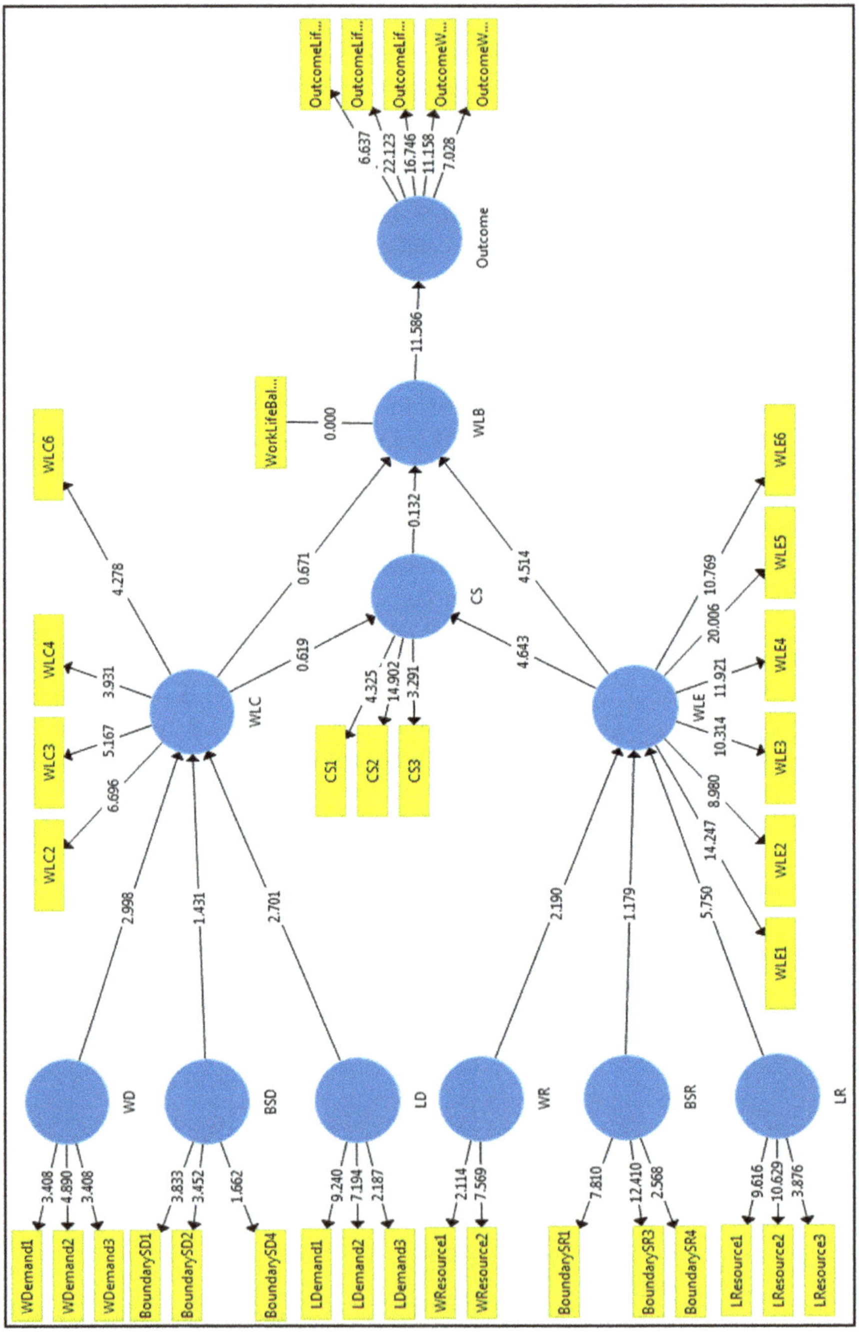

Fig. 6.16 Results of Bootstrapping *t*-values of the measurement and structural model relationships for South Korea

Table 6.42 Summary of the path coefficients, *t*-values and *p* values (South Korea)

	Path coefficients	*t*-values	*p* values	Significance ($p < 0.05$)?
BSD → WLC	0.188	1.431	0.152	No
BSR → WLE	0.120	1.179	0.238	No
CS → WLB	−0.029	0.132	0.895	No
LD → WLC	0.312	2.701	0.007	Yes
LR → WLE	0.592	5.750	0.000	Yes
WD → WLC	0.489	2.998	0.003	Yes
WLB → Outcome	0.756	11.586	0.000	Yes
WLC → CS	0.111	0.619	0.536	No
WLC → WLB	0.116	0.671	0.502	No
WLE → CS	0.566	4.643	0.000	Yes
WLE → WLB	0.605	4.514	0.000	Yes
WR → WLE	0.221	2.190	0.029	Yes

summarizes the structural model's path coefficients, *t*-values and *p* values. The significance and relevance of the structural model relationships would also be used to test the research hypotheses. The detailed evaluation relating to the research hypotheses would be explained later.

The total effect was evaluated to examine the indirect effects of other constructs on the endogenous latent variables. Table 6.43 summarizes the total effect values as well as results of the test of significance. Rows highlighted in grey were indirect constructs that have significant impact on other endogenous constructs. Based on Table 6.43, it could be seen that LR and WR which were antecedent of WLE have a significant total effect on CS, Outcome and WLB except for WD → Outcome which has a *p* value close to 0.05 (0.06). Thus, WR and LR played a significant role in describing the CS, WLB and Outcome (exclude WR) in the structural model. Paths highlighted in red do not pose any major effect on the model as the direct effect (path coefficient) and total effect were statistically significant.

The next procedure in evaluating the structural model used the R^2 value also known as the coefficient of determination. To reiterate, R^2 has a value ranging from 0 to 1 and currently there are no acceptable R^2 values as the result depends on the model complexity and the research objective. However, Hair et al. (2017) mentioned that R^2 values of 0.75, 0.50 and 0.25 for the endogenous constructs could be interpreted respectively as substantial, moderate and weak. Table 6.44 presents the R^2 values of all the endogenous latent variables. R^2 values for Outcome (0.572), WLC (0.574) and WLE (0.570) constructs could be considered to be moderate; and CS (0.384) and WLB (0.415) could be considered to be close to moderate.

Evaluating the effect size of f^2 was the next step in assessing the structural model. In addition to calculating the R^2 values of all the endogenous latent variables, the change in the R^2 value after the omission of one of its linked exogenous construct could be used to evaluate whether removing that construct would lead to a substantive impact on the endogenous latent variables (Hair et al., 2017). To

Table 6.43 Summary of test of significance results of the total effects (South Korea)

	Total Effects	t-values	*p* values	Significance ($p < 0.05$)?
BSD → CS	0.021	0.439	0.661	No
BSD → Outcome	0.016	0.454	0.650	No
BSD → WLB	0.021	0.456	0.648	No
BSD → WLC	0.188	1.431	0.152	No
BSR → CS	0.068	1.151	0.250	No
BSR → Outcome	0.054	1.062	0.288	No
BSR → WLB	0.071	1.094	0.274	No
BSR →WLE	0.120	1.179	0.238	No
CS → Outcome	-0.022	0.130	0.896	No
CS → WLB	-0.029	0.132	0.895	No
LD → CS	0.035	0.545	0.586	No
LD → Outcome	0.027	0.634	0.526	No
LD → WLB	0.035	0.623	0.533	No
LD → WLC	0.312	2.701	0.007	Yes
LR → CS	0.335	3.138	0.002	Yes
LR → Outcome	0.264	2.846	0.004	Yes
LR → WLB	0.349	3.165	0.002	Yes
LR → WLE	0.592	5.750	0.000	Yes
WD → CS	0.054	0.628	0.530	No
WD → Outcome	0.042	0.686	0.493	No
WD → WLB	0.055	0.695	0.487	No
WD → WLC	0.489	2.998	0.003	Yes
WLB → Outcome	0.756	11.586	0.000	Yes
WLC → CS	0.111	0.619	0.536	No
WLC → Outcome	0.085	0.699	0.485	No
WLC → WLB	0.113	0.699	0.484	No
WLE → CS	0.566	4.643	0.000	Yes
WLE → Outcome	0.445	3.562	0.000	Yes
WLE → WLB	0.589	4.113	0.000	Yes
WR → CS	0.138	1.996	0.046	Yes
WR → Outcome	0.098	1.883	0.060	No
WR → WLB	0.130	1.979	0.048	Yes
WR → WLE	0.221	2.190	0.029	Yes

reiterate, the rule of thumb in interpreting the f^2 values is as follows: 0.02 = small effects; 0.15 = medium effects; 0.35 = large effects; less than 0.02 = no effect (Hair et al., 2017). Following the rule of thumb as set out by Hair et al. (2017), the results from Table 6.45 could be interpreted as follows: BSD has small effects on WLC (0.077); BSR has small effect on WLE (0.024); CS has no effect on WLB (0.001); LD has medium effect on WLC (0.171); LR has large effect on WLE (0.561); WD has large effect on WLC (0.444); WLB has large effect on Outcome (1.334); WLC

Table 6.44 Summary of R^2 values on the structural model (South Korea)

	R^2
CS	0.384
Outcome	0.572
WLB	0.415
WLC	0.574
WLE	0.570

Table 6.45 Summary of f^2 effect size (South Korea)

f^2 Effect size	BSD	BSR	CS	LD	LR	OC	WD	WLB	WLC	WLE	WR
BSD									0.077		
BSR										0.024	
CS								0.001			
LD									0.171		
LR										0.561	
OC											
WD									0.444		
WLB						1.334					
WLC			0.017					0.019			
WLE			0.432					0.363			
WR										0.104	

has no effect on CS (0.017) and WLB (0.019); WLE has large effect on CS (0.432) and WLB (0.363); and WR has small to medium effect on WLE (0.104).

The fifth step in evaluating the structural model was to examine the predictive relevance of the Q^2 value. To reiterate, it has been posited that a Q^2 value of more than 0 for a specific reflective endogenous latent variable represents the path model's predictive relevancy for a dependent construct (Hair et al., 2017). The Q^2 value was obtained using the Blindfolding interface in SmartPLS. Table 6.46 presents the Construct Cross-validated Redundancy estimates. SSO represents the sum of the square observations (SSO) and SSE represents the sum of squared prediction errors (SSE). Q^2 was formulated to be 1-SSE/SSO and it was noted that only endogenous constructs have Q^2 values. Based on Table 6.46, it could be seen that all five endogenous constructs have Q^2 value well above zero (CS = 0.165, Outcome = 0.320, WLB = 0.314, WLC = 0.208 and WLE = 0.319). The results clearly supported the model's predictive relevancy regarding the endogenous latent variables.

The final step in evaluating the structural model was to assess the q^2 effect sizes. To reiterate, the use of q^2 is similar to the f^2 effect size approach for assessing R^2 where it assessed the change in the R^2 value after the omission of one of its linked exogenous construct. The effect size of q^2 also used this approach where it assessed an exogenous latent variable's contribution to the Q^2 value of an endogenous

Table 6.46 Summary of the Q^2 values (South Korea)

Construct cross-validated redundancy			
	SSO	SSE	Q^2 (1-SSE/SSO)
BSD	132.00	132.00	
BSR	132.00	132.00	
CS	132.00	110.17	0.165
LD	132.00	132.00	
LR	132.00	132.00	
OC	220.00	149.70	0.320
WD	132.00	132.00	
WLB	44.00	30.19	0.314
WLC	176.00	139.36	0.208
WLE	264.00	179.70	0.319
WR	88.00	88.00	

Table 6.47 Summary of the $Q^2_{excluded}$ and q^2 effect sizes (South Korea)

	$Q^2_{excluded}$	$Q^2_{included}$	q^2	Predictive relevance
BSD → WLC	0.202	0.208	0.008	Small
BSR → WLE	0.314	0.319	0.007	Small
CS → WLB	0.353	0.314	−0.057	Small
LD → WLC	0.184	0.208	0.030	Small
LR → WLE	0.178	0.319	0.207	Medium
WD → WLC	0.148	0.208	0.076	Small
WLC → CS	0.181	0.165	−0.019	Small
WLC → WLB	0.337	0.314	−0.034	Small
WLE → CS	0.056	0.165	0.131	~Medium*
WLE → WLB	0.081	0.314	0.340	~Large*
WR → WLE	0.299	0.319	0.029	Small

*~ denotes close to medium or large

construct. $Q^2_{included}$ shown in Table 6.47 was based on the Blindfolding estimation. The $Q^2_{excluded}$ value was obtained by manually deleting the corresponding predecessor of the endogenous latent variable and recalculating the Q^2 using the Blindfolding procedure (similar to the Step 5 approach). As an example WLE → WLB from Table 6.47 which summarized the $Q^2_{excluded}$ and q^2 Effect Sizes values, it could be seen that the path has a q^2 value of 0.340. This value was obtained by applying the formula ($q^2 = Q^2_{included} - Q^2_{excluded}/1 - Q^2_{included}$): (0.314 − 0.081)/(1 − 0.314). It is important to note that that 0.314 ($Q^2_{included}$) was the original Q^2 for the WLB construct retrieved from Step 5 when computing the Blindfolding procedure and 0.081 ($Q^2_{excluded}$) was retrieved by removing the WLE construct from the model and re-estimating the WLB using the Blindfolding

Table 6.48 Summary of the results for q^2 effect sizes

q^2 Effect sizes											
	BSD	BSR	CS	LD	LR	OC	WD	WLB	WLC	WLE	WR
BSD									0.008		
BSR										0.007	
CS								−0.057			
LD									0.030		
LR										0.207	
OC											
WD									0.076		
WLB						–					
WLC			−0.019					−0.034			
WLE			0.131					0.340			
WR										0.029	

procedure. To reiterate, Hair et al. (2017) mentioned that q^2 values of 0.02, 0.15 or 0.35 could be interpreted as the exogenous construct having a small, medium or large predictive relevance respectively on its linked endogenous latent variable. This also signified that the q^2 effect size for the relationship between WLE → WLB could be considered to be close to large predictive relevance (Table 6.48).

6.15 Overall Summary of the Inferential Statistics for Singapore and South Korea Result

To establish reliability and validity in the reflective measurement model, deletion of latent variables were inevitable. Table 6.49 lists the constructs removed from the model for both Singapore and South Korea. Once the model has established reliability and validity, the next step was to address the structural model results. A series of procedures were evaluated to examine the model's predictive

Table 6.49 List of indicators deleted to establish validity and reliability (Singapore and South Korea)

Deletion of indicators to establish validity and reliability		
	Singapore	South Korea
1	WDemand1	BoundarySD3
2	BoundarySD2	WResource3
3	BoundarySD3	BoundarySR2
4	LDemand1	WLC1
5	BoundarySR1	WLC5
6	CS3	CS4
7	OutcomeWork3	CS5
8		OutcomeWork3

Table 6.50 Summary of structural model's path coefficients, *p* values and conclusions for Singapore and South Korea results

	Singapore			South Korea		
	Path coefficients	*p* values	Conclusion (hypothesis)	Path coefficients	*p* values	Conclusions (hypotheses)
BSD → WLC	0.352	0.035*	Supported	0.188	0.152	Not Supported
BSR → WLE	−0.039	0.721	Not Supported	0.12	0.238	Not Supported
CS → WLB	0.561	0.032*	Supported	−0.029	0.895	Not Supported
LD → WLC	−0.088	0.487	Not Supported	0.312	0.007*	Supported
LR → WLE	0.527	0.000*	Supported	0.592	0.000*	Supported
WD → WLC	0.555	0.001*	Supported	0.489	0.003*	Supported
WLB → Outcome	0.554	0.000*	Supported	0.756	0.000*	Supported
WLC → CS	0.072	0.476	Not Supported	0.111	0.536	Not Supported
WLC → WLB	0.161	0.310	Not Supported	0.116	0.502	Not Supported
WLE → CS	0.776	0.000*	Supported	0.566	0.000*	Supported
WLE → WLB	0.006	0.812	Not Supported	0.605	0.000*	Supported
WR → WLE	0.387	0.001*	Supported	0.221	0.029*	Supported

*Path significance: $p < 0.05$

Table 6.51 Summary of the outer loading's *t*-values and *p* values for Singapore's results

Observed variables	Latent variables	Outer loading	*t*-values	*p* values	Significance ($p < 0.05$)?
WDemand2	WD	0.759	7.576	0.000	Yes
WDemand3		0.862	10.016	0.000	Yes
BoundarySD1	BSD	0.620	2.837	0.005	Yes
BoundarySD4		0.934	26.983	0.000	Yes
LDemand2	LD	0.989	2.854	0.004	Yes
LDemand3		0.982	2.891	0.004	Yes
WResource1	WR	0.673	3.910	0.000	Yes
WResource2		0.728	4.087	0.000	Yes
WResource3		0.775	4.400	0.000	Yes
BoundarySR2	BSR	0.911	4.917	0.000	Yes
BoundarySR3		0.797	5.654	0.000	Yes
BoundarySR4		0.900	4.451	0.000	Yes
LResource1	LR	0.729	5.942	0.000	Yes
LResource2		0.857	14.184	0.000	Yes
LResource3		0.776	7.232	0.000	Yes
WLC1	WLC	0.761	6.699	0.000	Yes
WLC2		0.738	5.607	0.000	Yes
WLC3		0.855	22.680	0.000	Yes
WLC4		0.758	6.304	0.000	Yes
WLC5		0.673	4.210	0.000	Yes
WLC6		0.855	13.102	0.000	Yes
WLE1	WLE	0.716	5.122	0.000	Yes
WLE2		0.780	8.028	0.000	Yes
WLE3		0.726	5.534	0.000	Yes
WLE4		0.738	8.967	0.000	Yes
WLE5		0.821	13.552	0.000	Yes
WLE6		0.834	14.092	0.000	Yes
CS1	CS	0.829	12.246	0.000	Yes
CS2		0.849	9.917	0.000	Yes
CS4		0.730	5.829	0.000	Yes
CS5		0.721	6.015	0.000	Yes
OutcomeLife1	Outcome	0.663	3.129	0.002	Yes
OutcomeLife2		0.657	4.521	0.000	Yes
OutcomeLife3		0.854	11.233	0.000	Yes
OutcomeWork1		0.708	3.013	0.003	Yes
OutcomeWork2		0.756	3.877	0.000	Yes

Table 6.52 Summary of the outer loading's *t*-values and *p* values for South Korea result

Observed Variables	Latent Variables	Outer Loading	*t-values*	*p* values	Significance ($p < 0.05$)?
WDemand1	WD	0.622	3.355	0.001	Yes
WDemand2		0.822	5.855	0.000	Yes
WDemand3		0.755	3.806	0.000	Yes
BoundarySD1	BSD	0.888	3.476	0.001	Yes
BoundarySD2		0.845	3.214	0.001	Yes
BoundarySD4		0.511	1.732	0.084	No
LDemand1	LD	0.854	10.704	0.000	Yes
LDemand2		0.833	7.193	0.000	Yes
LDemand3		0.512	2.241	0.025	Yes
WResource1	WR	0.597	2.091	0.037	Yes
WResource2		0.907	8.528	0.000	Yes
BoundarySR1	BSR	0.860	8.330	0.000	Yes
BoundarySR3		0.921	13.899	0.000	Yes
BoundarySR4		0.564	2.729	0.006	Yes
LResource1	LR	0.791	10.204	0.000	Yes
LResource2		0.839	10.634	0.000	Yes
LResource3		0.649	4.019	0.000	Yes
WLC2	WLC	0.802	5.584	0.000	Yes
WLC3		0.707	4.289	0.000	Yes
WLC4		0.723	4.153	0.000	Yes
WLC6		0.672	3.566	0.000	Yes
WLE1	WLE	0.854	14.504	0.000	Yes
WLE2		0.751	9.809	0.000	Yes
WLE3		0.781	10.622	0.000	Yes
WLE4		0.803	12.557	0.000	Yes
WLE5		0.842	20.931	0.000	Yes
WLE6		0.794	11.240	0.000	Yes
CS1	CS	0.679	4.098	0.000	Yes
CS2		0.866	12.502	0.000	Yes
CS3		0.731	3.484	0.001	Yes
OutcomeLife1	Outcome	0.737	6.648	0.000	Yes
OutcomeLife2		0.861	22.274	0.000	Yes
OutcomeLife3		0.833	16.752	0.000	Yes
OutcomeWork1		0.770	11.504	0.000	Yes
OutcomeWork2		0.775	6.971	0.000	Yes

capabilities and the relationships between the endogenous and exogenous constructs. Table 6.50 shows the path coefficients and the *p* values of all the relationships between constructs. In addition, the Conclusion column indicates whether the hypotheses formulated for the research were or were not supported.

Tables 6.51 and 6.52 summarize the outer loading's *t*-values and *p* values for Singapore and South Korea's results respectively. Both results were obtained using another round of Bootstrapping procedure. Since Bootstrapping is a nonparametric procedure, the *t*-values and *p* values were different from the first round of Bootstrapping. However, the difference was not significant and did not affect the results substantially. Since this was a reflective measurement model, the causal relationship was from the latent variable to its observed variable. Singapore's results in Table 6.51 showed that the entire latent variables were significant in explaining the observed variables. South Korea's results in Table 6.52 however, has one observed variable (BoundarySD4) which has a *p* value more than 0.05 (0.084). This means the latent variable (BSD) was not significant in explaining BoundarySD4 and hence it was rejected. Other than BoundarySD4, the rest of the outer loadings were significant (<0.05). This also implied that the causality from the construct (latent variables) to its indicators (observed) was significant.

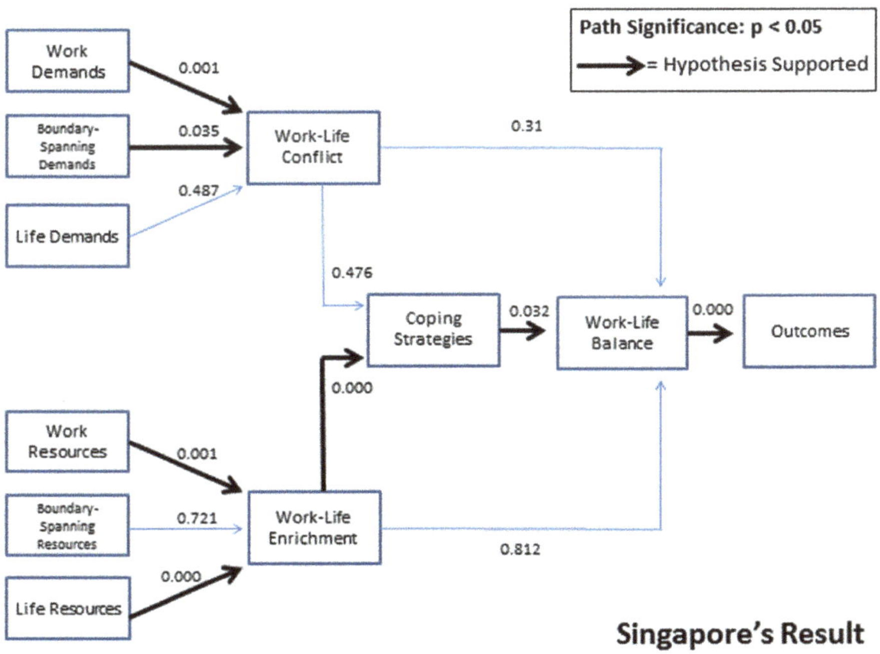

Fig. 6.17 Path significance of the structural model (Singapore)

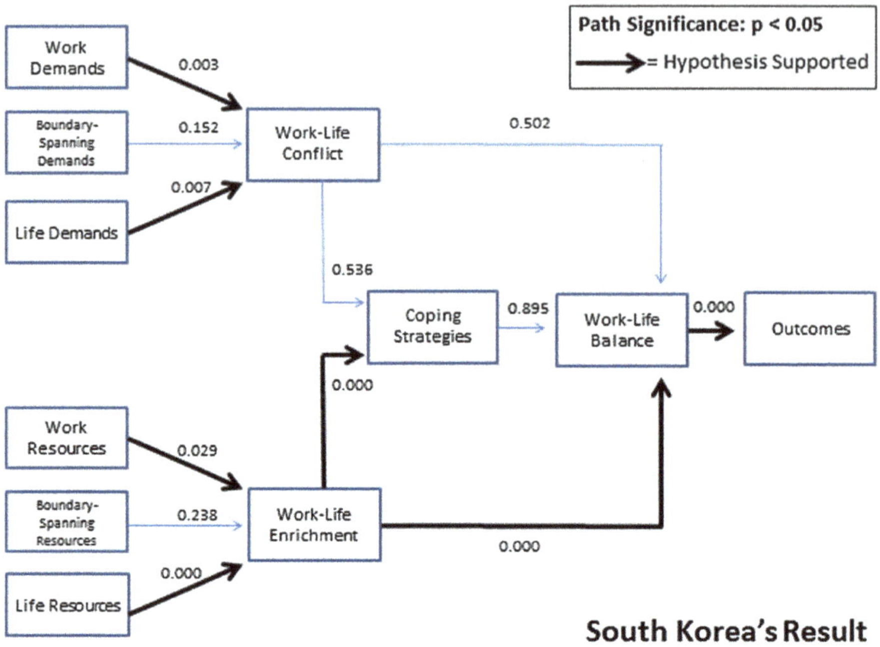

Fig. 6.18 Path significance of the structural model (South Korea)

6.16 Summary of Results and Findings

6.16.1 Overview

Detailed explanation of the results and findings were presented in this chapter. Firstly, the hypotheses formulated in this study were compared against the results obtained from the path analysis. Discussions of the results for each hypothesis for Singapore and South Korea were analyzed separately to understand the reasons for their significant or non-significant relationships. Verbatim transcripts collated from interviews were also used to verify the result. Figures 6.17 and 6.18 show the overview of the path coefficient of the structural model for Singapore and South Korea respectively.

6.16.2 Discussions of Results and Interview Findings for Singapore

Hypothesis 1a (H1a) states that there is a positive relationship between Work Demands and Work–Life Conflict. In Singapore's findings, the hypothesized path for H1a was positive and significant (path coefficient = 0.555, $p = 0.001$). Thus, the hypothesized H1a was supported. The result supports the model proposed by Voydanoff (2005a) and Carlson et al. (2000) where Work–Life Conflict was the cognitive appraisal of the effects of work demands (antecedent). In alignment with the results, both Miss A and Mr B agreed that Work Demand was important in explaining Work–Life Conflict. As mentioned by Miss A:

> *Because is how much your workload and your amount of responsibilities that determines how much stress you are under.*

Mr B also mentioned:

> *I think everyone will encounter work demand in their career is just the degree of impact it has on the individual conflict.*

High work demand was unavoidable especially in the construction industry due to the nature of construction projects (tight deadlines, high workload, etc.). Miss A and Mr B found that among the three demands (work, boundary spanning and life), Work Demands was more likely to cause Work–Life Conflict.

Hypothesis 1b (H1b) states that there is positive relationship between Boundary Spanning Demands and Work–Life Conflict. In Singapore's findings, the hypothesized path for H1b was positive and significant (path coefficient = 0.352, $p = 0.035$). Thus, the hypothesized H1a was supported. The result supports the model proposed by Voydanoff (2005b) where Boundary Spanning Demands encompassed both Work and Life Demand even though it could originate in either one domain. Mr B agreed that Boundary Spanning Demand was important in explaining Work–Life Conflict. As explained by Mr B:

> *I think it will have impact especially the amount of over time (OT) we have.*

OT or overtime in this case referred to the workload that happened when the project team was unable to complete the tasks within the normal working hour and have to result in working beyond normal work hours. Consequently, the time spent in life domain would be reduced resulting in work and life conflict. However, Miss A personally felt less so because the policies in her company restricted work from affecting the life domain:

> *Usually policies that are implemented in my organization are usually with positive intention less of a Boundary Spanning Demand.*

The two interviewees were from different organizations and it could be seen that the policies implemented by their organizations have an adverse impact in their WLB not just Boundary Spanning Demand. A policy that restricted employees from working overtime would help the individual reduce Work–Life Conflict.

Hypothesis 1c (H1c) states that there is positive relationship between Life Demands and Work–Life Conflict. In Singapore's findings, the hypothesized path for H1a was not significant (path coefficient = −0.088, $p = 0.487$). Thus, the hypothesized H1c was not supported. Similar to the result, Miss A disagreed that

Life Demand is important in explaining Work–Life Conflict. As explained by Miss A:

> *Work is still a primary consideration and life demands don't really affect much of my Work–Life Conflict.*

Miss A found that the Life Demand perspective might change as the responsibilities in life increase (e.g. married or having children). Mr B agreed that Life Demand is important in explaining Work–Life Conflict and added that:

> *Same as work demand is the matter of degree of impact on the individual. Different people will have different impact at certain times.*

However, of all three demands, Mr B surmised that Work Demand would have a major impact because of the amount of work volume at times which would affect the life domain. An example was given by Mr B:

> *One of my staff wants to apply three weeks leave but I only can grant him two weeks leave due to the amount of work load she has. I understand it may cause dissatisfaction but no choice as three weeks is too long*

Hypothesis 2a (H2a) states that there is positive relationship between Work Resources and Work–Life Enrichment. In Singapore's findings, the hypothesized path for H2a was positive and significant (path coefficient = 0.387, $p = 0.001$). Thus, the hypothesized H2a was supported. The results supported the model proposed by Carlson et al. (2006) where they defined the acquisition and the use of resources was critical to the occurrence of Work–Life Enrichment. Likewise, the application of resources would help to facilitate performance, reduce demands and/or generate additional resources to an individual (Voydanoff, 2005b). Mr B agreed that Work Resources is important in explaining Work–Life Enrichment:

> *Training resources provided in the company also help the employees upgrade their skills and knowledge and benefit them to progress further.*

Miss A however does not find Work Resources linked to Work–Life Enrichment but rather a cause of job satisfaction. She commented that:

> *I think Work Resources is more of a job satisfaction than work life balance.*

Hypothesis 2b (H2b) states that there is positive relationship between Boundary Resources and Work–Life Enrichment. In Singapore's findings, the hypothesized path for H2b was not significant (path coefficient = −0.039, $p = 0.721$). Thus, the hypothesized H2b was not supported. Mr B agreed that Boundary Spanning Resource was important in explaining Work–Life Enrichment. His organization occasionally organized events that also allowed the employee's family to join in. As explained by Mr B:

> *Our company will try to provide such resources annually like for example family day or dinner and dance event. During the last family day, we went to the zoo and all the*

> *employees will bring their loved ones along. This is a good way to allow employee's family to interact with their colleagues or bosses.*

Miss A however mentioned that her organization does not provide such resources and that:

> *I think is important but my company does not really provide such resources at the moment as you know construction projects are hard to allow team member to work from home as client may require him to be present when he needs you.*

Similar to Boundary Spanning Demand, policies implemented by the organization played a part in the level of Boundary Spanning across work and life domain.

Hypothesis 2c (H2c) states that there is positive relationship between Life Resources and Work–Life Enrichment. In Singapore's findings, the hypothesized path for H2c was positive and significant (path coefficient = 0.527, $p = 0.000$). Thus, the hypothesized H2c was supported. The results supported the model proposed by Carlson et al. (2006) where they predicted that enrichment would be positively correlated with life antecedents. Likewise, the application of resources would help to facilitate performance, reduce demands and/or generate additional resources to an individual (Voydanoff, 2005b). Both Miss A and Mr B agreed that Life Resources was important in explaining Work–Life Enrichment. Miss A mentioned that Life Resources was one way that would make her feel motivated and to recuperate from high stresses at work. She explained that:

> *Sometimes weekend can help let you look forward to your following weekday. Sometimes when your weekend is ruined, it may make my weekday moody. Psychological reward is a very good way in helping me feel motivated. And also when you have high work demand, sometimes spend time with friends and family can help to recuperate from the stresses.*

Mr B felt that life resources could help Singaporeans compensate for the demand as opposed to the foreign workers in the construction industry who could not rely on their families or loved ones whenever they needed them. He explained that:

> *Foreign talent in Singapore may be less likely as they do not have their loved ones here. For Singaporeans, it will definitely as the life resources can help to compensate the demand that the employee currently has.*

Hypothesis 3a (H3a) states that there is positive relationship between Work–Life Conflict and Coping Strategies. In Singapore's findings, the hypothesized path for H3a was not significant (path coefficient = 0.072, $p = 0.476$). Thus, the hypothesized H3a was not supported. However, Miss A and Mr B felt there was a degree to which Work–Life Conflict was associated with Coping Strategies. Miss A mentioned that work demand in her assignment could be substantial at times and coping strategy was definitely required. Of the five Coping Strategies proposed in this present study, she felt that cognitive management and behavioral management were the two coping strategies needed to help her cope with high Work–Life Conflict. She commented that:

For people who OT a lot, I feel that they require cognitive management more as they are used to it and be still able to attain WLB. This happens to my colleagues as well because they OT a lot but they don't feel the negative effect at all and still be able to handle their non-work schedule.

Mr B found that even though Work–Life Conflict was not significant in explaining Coping Strategies, he still found there was a degree to which people would need it to reduce the Work–Life Conflict. Mr B explained that:

I find that work life conflict cannot be chunked aside as I felt that it will have an adverse effect on an individual work life balance. That is why you need coping strategies to cope with the conflict.

However, Miss A made a proposition to explain the reason why the hypothesis was not significant. Miss A added that:

I feel that there is certain limit as to how much people will have in their Work–Life Conflict once it reaches the threshold, I don't think Work–Life Enrichment will be beneficial to explain Coping Strategies. Reduce the work life conflict then you can think of the work life enrichment. Maybe your survey of all the respondents have not reached the threshold that is why work life enrichment is significant.

Miss A felt there was a limit to which an individual could tolerate the Work–Life Conflict. Once exceeded, coping strategies would definitely be needed. This could also suggest why the respondents felt that the level of Work–Life Conflict was not substantial enough to trigger the Coping Strategies.

Hypothesis 3b (H3b) states that there is positive relationship between Work–Life Enrichment and Coping Strategies. In Singapore's findings, the hypothesized path for H3a was positive and significant (path coefficient = 0.776, $p = 0.000$). Thus, hypothesized H3b was supported. The results concurred with the model proposed by Haddon and Hede (2009) where they posited that even if low enrichment was a likely consequence of engaging in work and life roles, it was how people cope with the situation that determined their outcomes. Both Miss A and Mr B agreed that Coping Strategies were needed to alter the outcome of Work–Life Enrichment. As Miss A mentioned:

I feel that it is the supports and leisure activities that help you improve your work life enrichment. But that doesn't mean the rest are not important; it is just that it may not contribute substantially.

Mr B felt that Work–Life Enrichment is important in overcoming the negative effect of Work–Life Conflict. As Mr B explained:

Reducing conflict itself is not enough in my view but with the enrichment available, the Millennials can be more likely to achieve work life balance. Maybe it is because in the construction industry there may not be a lot of enrichment available which is why coping strategies is needed to overcome the low enrichment.

His view was that Millennials employees might not have achieved the level of Work–Life Enrichment needed; hence, the reason why Coping Strategies were needed to enhance Work–Life Enrichment. Furthermore, Mr B emphasized that work life enrichment must always be available to the individual:

> *Enrichment must always be available to the individual and we must try to provide enhancement strategies to help them increase their level of enrichment.*

Hypothesis 4a (H4a) states that there is negative relationship between Work–Life Conflict and Work–Life Balance. In Singapore's findings, the hypothesized path for H4a was not significant (path coefficient = 0.161, $p = 0.310$). Thus, the hypothesized H4a was not supported. Mr B agreed that Work–Life Conflict was not associated with WLB in this study because:

> *I felt that even though workers may encounter major conflict in their work, if enrichment is present it can help to overshadow the high conflict which is why conflict is not associated to work life balance in your study. I agree to your explanation and will like to add on. Reducing conflict itself is not enough in my view but with the enrichment available, the Millennials can be more likely to achieve work life balance.*

Mr B felt that Work–Life Enrichment is the key forward no matter how much Work–Life Conflict an individual might encounter; high Work–Life Enrichment must be present to overshadow Work–Life Conflict. Miss A however, disagreed that Work–Life Conflict was not able to explain WLB. She commented that:

> *I think for blue collar, low work life conflict is enough for them to attain WLB as they don't expect too much from their job. White collar then we must factor in high work–life enrichment. For me I definitely feel that both contribute to higher level of Work–Life Balance.*

Hypothesis 4b (H4b) states that there is positive relationship between Work–Life Enrichment and Work–Life Balance. In Singapore's findings, the hypothesized path for H4b was not significant (path coefficient = 0.006, $p = 0.812$). Thus, the hypothesized H4b was not supported. However, the total effect between Work–Life Enrichment and Work–Life Balance was significant (Total Effect = 0.493, $p = 0.000$), which means that the indirect effect between the two latent variables was significant but not the direct effect. The findings provided empirical support for the mediating role of Coping Strategies. More specifically, Coping Strategies were seen as a mechanism underlying the relationship between Work–Life Enrichment and Work–Life Balance. Work–Life Enrichment must be mediated by Coping Strategies for the Singaporeans to achieve Work–Life Balance. Both Miss A and Mr B felt that Work–Life Enrichment was important in helping the individual achieve WLB. As mentioned previously, Miss A has commented that both Work–Life Conflict and Work–Life Enrichment contributed to higher level of WLB. Mr B added that due to the nature of the construction industry, it might be difficult for the targeted respondents to have high level of Work–Life Enrichment. That was why Coping Strategies were needed to overcome the low Work–Life Enrichment.

Hypothesis 4c (H4c) states that there is positive relationship between Coping Strategies and Work–Life Balance. In Singapore's finding, the hypothesized path for H4c was positive and significant (path coefficient = 0.561, p = 0.032). Thus, the hypothesized H4c was supported. The results supported the model proposed by Haddon and Hede (2009) where they posited that research pertaining to Work–Life Balance must focus on the exploration of relevant coping strategies which directly led to the achievement of Work–Life Balance. Both Miss A and Mr B agreed that Coping Strategies were important in explaining WLB. Miss A felt that the construction industry was tough and demanding and Coping Strategies must be present to help employees achieve WLB. As explained by Miss A:

> *Especially the expectation in my job if I have a high expectation in my work, coping strategy will be important in helping me cope with the negative effect in order for me to get my job done.*

Similarly, Mr B added that:

> Coping strategies must also be present to help contribute to work life balance of an individual.

Hypothesis 5 (H5) states that there is positive relationship between Work–Life Balance and Outcomes. In Singapore's findings, the hypothesized path for H5 was positive and significant (path coefficient = 0.554, p = 0.000). Thus, the hypothesized H5 was supported. The results of this significance supported many past research studies that affirmed achieving Work–Life Balance would yield beneficial consequences in both work and life domain (Carlson et al., 2009; Grzywacz & Carlson, 2007; Haddon & Hede, 2010). Both Miss A and Mr B agreed that high WLB would lead to positive outcomes. The interviewees were also asked for their views on each of the outcomes adopted in this study: job satisfaction, organizational commitment, intention to turnover, family satisfaction, life satisfaction and psychological well-being. Both Miss A and Mr B agreed that WLB would lead to job, life and family satisfaction. As for intention for turnover, both felt otherwise. Miss A commented that:

> *Turnover intention is more on opportunity if given an opportunity higher salary and career advancement I will quit my job immediately but because of WLB I stay in my job is no for me.*

Likewise, Mr B commented that:

> *As the job here is very demanding and the pay is below or average market rate, given an opportunity (good pay or more career advancement) they will leave the company no doubt. Work life balance itself is not enough to retain them.*

It seems that both interviewees felt that having WLB does not necessary reduce the turnover intention of Singaporean Millennials who might leave given better opportunities elsewhere. This also concurred with the findings where intention to turnover was unreliable and invalid in explaining the Outcomes. Likewise, Miss A disagreed WLB would lead to organization commitment and psychological well-being. She commented that:

For organizational commitment, is more on people related factors like relationships with colleagues and bosses that lead to your commitment. Psychological well-being I don't feel it now, it does not apply to me now.

Mr B on the other hand agreed that WLB would lead to organizational commitment and psychological well-being.

6.16.3 Discussions of Results and Interview Findings for South Korea

Hypothesis 1a (H1a) states that there is positive relationship between Work Demands and Work–Life Conflict. In South Korea's findings, the hypothesized path for H1a was positive and significant (path coefficient = 0.489, $p = 0.003$). Thus, the hypothesized H1a was supported. The result supported the model proposed by Voydanoff (2005a) and Carlson et al. (2000) where Work–Life Conflict was the cognitive appraisal of the effects of work demands (antecedent). The two interviewees Mr C and Mr D also agreed that Work Demand was important in explaining Work–Life. As explained by Mr C:

I have already worked in this company for 10 years and my position is getting higher. My job requirements also increase. My working scope has increased and I have to take care of more people and activities. That means I cannot be free from work even I go back home. They don't care whether I end work at 2 am or 3 am because it is my duty to take care of the overall project. The more work I received, the more stressed I am. This is where conflict part comes in.

The increasing job responsibilities due to his senior designation has made Mr C's work demand increased significantly. Mr D mentioned that:

Work is always the main source of problem regardless how great your life is. Work demand causes work and life conflicting with each other no doubt.

Moreover, Mr C and Mr D found that among the three demands, Work Demands is more likely to cause Work–Life Conflict.

Hypothesis 1b (H1b) states that there is positive relationship between Boundary Spanning Demands and Work–Life Conflict. In South Korea's findings, the hypothesized path for H1b was not significant (path coefficient = 0.188, $p = 0.152$). Thus, the hypothesized H1b was not supported. However, both Mr C and Mr D agreed that Boundary Spanning Demands do have effect on Work–Life Conflict. Mr C noted that:

Just like I explained to you, even I go back home without OT, without OT payment I still have to work. Sometimes I have work over the holiday and Saturday or even some Sundays without any OT payment. Our company does not give us OT payment. Even I go back home I cannot be free from work.

Mr D similarly mentioned that:

> *You cannot avoid overtime work especially during peak period where work has to be done as soon as possible. Bringing work back home is very common for us not just for the South Koreans.*

Overtime work was highlighted by two interviewees as one of the main reasons affecting both work and life domain which eventually caused an increase in Work–Life Conflict.

Hypothesis 1c (H1c) states that there is positive relationship between Life Demands and Work–Life Conflict. In South Korea's findings, the hypothesized path for H1c was positive and significant (path coefficient = 0.312, $p = 0.007$). Thus, the hypothesized H1c was supported. The results supported the model proposed by Voydanoff (2005a) and Carlson et al. (2000) where Work–Life Conflict was the cognitive appraisal of the effects of life demands (antecedent). Mr C agreed that Life Demands was important in explaining Work–Life Conflict. As Mr C explained,

> *I am a father and I am a husband. I have the duty to support the family. I also have the duty for this project. My wife will complain why I still work even I come back home, I usually explain to her but she sometime don't understand.*

Mr D however found that Life Demands does not affect the South Korean's Work–Life Conflict substantially, as he mentioned that:

> *They do not have friends or family here so I think life demand may not be so impactful towards the work life conflict.*

Hypothesis 2a (H2a) states that there is positive relationship between Work Resources and Work–Life Enrichment. In South Korea's findings, the hypothesized path for H2a was positive and significant (path coefficient = 0.221, $p = 0.029$). Thus, the hypothesized H2a was supported. The result supported the model proposed by Carlson et al. (2006) where they defined the acquisition and the use of resources to be critical to the occurrence of Work–Life Enrichment. Likewise, the application of resources would help to facilitate performance, reduce demands and/or generate additional resources to an individual (Voydanoff, 2005b). Both Mr C and Mr D agreed that Work Resources were important in explaining Work–Life Enrichment. Mr C felt that promotion and salary were the key reasons that enhanced his work and life domain when he noted that:

> *Promotion and increasing salary is the reason I came here and this one of course if I can earn more money I can feel motivated. Sometime I also feel achievement when I settled a challenging work.*

Mr D also commented that:

> *The South Koreans I met are all very hardworking and during my conversation with one of the staff over lunch break, she told me that stationing in Singapore can help them progress.*

This suggested that career progression might act as a form of resource which helped to increase Work–Life Enrichment for the individuals.

Hypothesis 2b (H2b) states that there is positive relationship between Boundary Spanning Resources and Work–Life Enrichment. In South Korea's findings, the hypothesized path for H2b was not significant (path coefficient = 0.12, p = 0.238). Thus, the hypothesized H2b was not supported. Both Mr C and Mr D disagreed that Boundary Spanning Resources were not important in explaining Work–Life Enrichment. Mr C commented that these resources were not implemented in his company for several reasons:

> *I currently don't have flexible work schedule or work at home. As you know, South Korean culture is very rigid not flexible. Our hierarchy is also very strong and our working hours are from 7 am to 6 pm. Construction team had to work 24 h around the clock because we have to think about the actual site conditions. Have to monitor the site.*

Mr D also felt that Boundary Spanning Resources does not help much in enriching individual's work and life domain when he commented that:

> *Enticing the staffs by giving them time off to do their other stuff I don't think can help enrich their personal life. The thing you said like annual dinner and dance we do have but once a year I don't think can help them much. If you say flexible work arrangement, yes it will help but I don't allow it because you may never know when I need them.*

Hypothesis 2c (H2c) states that there is positive relationship between Life Resources and Work–Life Enrichment. In South Korea's findings, the hypothesized path for H2c was positive and significant (path coefficient = 0.592, p = 0.000). Thus, the hypothesized H2c was supported. The results supported the model proposed by Carlson et al. (2006) where they predicted enrichment would be positively correlated with life antecedents. Likewise, the application of resources would help to facilitate performance, reduce demands and/or generate additional resources to an individual (Voydanoff, 2005b). Mr C agreed that Life Resources was important in explaining individual's Work–Life Enrichment when he commented that:

> *Actually just six months ago I brought my family here. Before that we live separately. When we live together, I feel very happy. When I go back home play with my daughter even though 10 minutes or 15 minutes I can smile. I work for the sake of my family. One time I can feel stress but when I see my daughter I will feel happy.*

Mr D on the other hand felt that South Koreans in Singapore do not have much Life Resources available as their loved ones were not here for them but perspective would likely change if their loved ones were here for them. He explained that:

> *Of course it is important but not that much for them because they don't have friends and families here. Assume that they are working in South Korea, then I would say that life resources is definitely important in enriching their work and life.*

Hypothesis 3a (H3a) states that there is positive relationship between Work–Life Conflict and Coping Strategies. In South Korea's findings, the hypothesized path for H3a was not significant (path coefficient = 0.111, p = 0.536). Thus, the hypothesized H3a was not supported. Mr C however felt that Coping Strategy was definitely needed when a person experienced high Work–Life Conflict and this developed when an individual started to adapt to the working environment and learned to cope with issues relating to their work. He added that:

> *If you every day faces with high demand causing work life conflict you cannot expect the conflict to stay with you throughout the week. Like I said earlier, the South Koreans may face issue adapting to our culture but as times goes by they started to blend in. This is the same logic.*

Mr C was however unable to provide any comment relating to this hypothesis.

Hypothesis 3b (H3b) states that there is positive relationship between Work–Life Enrichment and Coping Strategies. In South Korea's findings, the hypothesized path for H3a was positive and significant (path coefficient = 0.566, p = 0.000). Thus, hypothesized H3b was supported. The results concurred with the model proposed by Haddon and Hede (2009) where they posited that even if low enrichment was a likely consequence of engaging in work and life roles, it was how people coped with the situation that determine their outcomes. Mr D felt that everyone needed Coping Strategies at work or personal life when he mentioned that:

> *Coping Strategy is needed to meet your Work–Life Balance. You cannot expect a guy without any coping strategy to answer you that he has work life balance. Everyone will have their own coping strategy but maybe different from each other.*

Mr C also agreed to this hypothesis but only to a certain extent. He felt that Coping Strategies is temporary when he commented that:

> *Having coping strategies is good but it is only temporary. Maybe you go overseas relax they still need to work. For example, before my family came here I would have traveled back to Korea every 4 months to visit them. This regular holiday can be a coping strategy. When I go back Korea I can relax without work pressure but after 1 week I will start to be concerned about my work. Because when I go back to work everyone will wait for me.*

Hypothesis 4a (H4a) states that there is negative relationship between Work–Life Conflict and Work–Life Balance. In South Korea's findings, the hypothesized path for H4a was not significant (path coefficient = 0.116, p = 0.502). Thus, the hypothesized H4a was not supported. Mr D felt that Work–Life Conflict was not significant in explaining WLB is because individuals already knew what they would be facing in the construction industry when they chose to enter this industry due to other (positive) reasons that might help them attain WLB. Mr D commented that:

> *Low conflict is not significant maybe is because the South Koreans have already knew that in construction you will definitely face with conflict of interest between work and life. They choose to work in construction means they knew they will face these consequences and willing to make this sacrifice.*

Mr C on the other hand mentioned that:

> *Work life conflict we can control one time we adjust it, it can become normal. If they earn more money they can bear more stress.*

This suggests that the availability of Work–Life Enrichment has helped to overcome the Work–Life Conflict of an individual easily, in this case Mr C felt that monetary rewards helped to overshadow the high Work–Life Conflict in his work.

Hypothesis 4b (H4b) states that there is positive relationship between Work–Life Enrichment and Work–Life Balance. In South Korea's findings, the hypothesized path for H4b was positive and significant (path coefficient = 0.605, $p = 0.000$). Thus, hypothesized H4b was supported. The results supported the model proposed by Grzywacz and Carlson (2007), Haddon and Hede (2010) and Rantanen, Kinnunen, Maunom, and Tillemann (2010) where they posited that high Work–Life Enrichment and low Work–Life Conflict would result in the achievement of WLB. However, since Work–Life Enrichment and Work–Life Conflict reflected orthogonal dimensions, it was possible that the presence of either one of the dimensions would help individuals achieve WLB (Rantanen et al., 2010). In this case, the South Korean respondents found that Work–Life Enrichment has a stronger effect in explaining WLB. Mr C agreed that Work–Life Enrichment was important in explaining WLB when he explained that:

> *Actually for me Work Life Enrichment for us is increase in salary and to make more money. In Singapore and Korea I think money is very important for them and us to survive.*

Mr C deemed Work–Life Enrichment as a form of monetary incentives and felt that it was the reason why he thought he was still able to attain WLB. Mr D deemed Work–Life Enrichment as a form of career advancement when he commented that:

> *High enrichment as a form of career progression for them is desirable especially for those who are willing to sacrifice their time with friends and families and work in Singapore.*

Hypothesis 4c (H4c) states that there is positive relationship between Coping Strategies and Work Life Balance. In South Korea's findings, the hypothesized path for H4c was not significant (path coefficient = −0.029, $p = 0.895$). Thus, the hypothesized H4c was not supported. Mr C felt that Coping Strategies was a temporary solution and might be less effective in the achievement of WLB. He noted that:

> *Of course it will help even though I find that it is temporary. Even though your result indicates that it is not significant may be is because it is less effective.*

Mr D however argued that Coping Strategies was needed to attain WLB. To reiterate, he commented that:

> *Coping Strategy is needed to meet your WLB. You cannot expect a guy without any coping strategy to answer you that he has work life balance. Everyone will have their own coping strategy but maybe different from each other.*

Hypothesis 5 (H5) states that there is positive relationship between Work–Life Balance and Outcomes. In South Korea's findings, the hypothesized path for H5 was positive and significant (path coefficient = 0.756, $p = 0.000$). Thus, the hypothesized H5 was supported. The results of this significance supported many past research studies that reported achieving WLB would yield beneficial consequences in both work and life domain (Carlson et al., 2009; Grzywacz & Carlson, 2007; Haddon & Hede, 2010). Both Mr C and Mr D agreed that achieving WLB would lead to positive outcomes. Mr C mentioned that:

> *Work–Life Balance will definitely help in my family satisfaction. Work also if I have work life balance I will have less stress which help me be more effective in my work and make me become more happy.*

Mr D felt that family satisfaction might not be relevant for the South Koreans who are working in Singapore.

6.17 Summary of the Results from Singapore and South Korea

Both Singapore and South Korea's results showed a similarity in explaining WLB based on the conceptual framework. Figures 6.11 and 6.12 showed that Work Resources and Life Resources were seen as the antecedents that contributed significantly in explaining Work–Life Enrichment in both Singapore and South Korea results. Furthermore, the total effects from both results also suggested that work and life resources played a significant role in the explanation of WLB where the indirect effects between work and life resources to WLB was statistically significant. The findings concurred with Haddon and Hede's (2010) study where their empirical evidence also suggested that work and life resources have a positive impact on an individual's Work–Life Enrichment and WLB. Both results indicated that Work–Life Enrichment was significant in explaining WLB be it a direct relationship (South Korea) or an indirect relationship (Singapore (Coping Strategies as mediator)). However, based on the literature review, it was observed that high Work–Life Enrichment and low Work–Life Conflict should be present for individuals to attain WLB (Frone, 2003; Grzywacz & Carlson, 2007; Haddon & Hede, 2010). Despite the differences, the results suggested that Singaporeans and South Koreans required high Work–Life Enrichment in order to attain their WLB even with high or low Work–Life Conflict. This could be explained by Herzberg's two-factor theory where there were two different sets of needs and where the different types of work environment satisfied and dissatisfied these needs. Hygiene factors were seen as basic needs (salary, job security) that individuals must achieve in order to reduce their dissatisfaction, but it only prevented dissatisfaction (Lundberg, Gudmundson, & Andersson, 2009). Motivators on the other hand were concerned with the growth needs (career advancement, development opportunities) and achieving these would deliver satisfaction and enhance work or life motivation (Lundberg et al., 2009). Similar to Work–Life Enrichment and Work–Life Conflict, Hygiene factors and motivators were independent constructs and have distinct sets of indicators. Applying the motivators do not remedy hygiene factors and vice versa. It was only when hygiene factors were fulfilled that workers were neither satisfied nor dissatisfied and the presence of motivators would make them motivated and satisfied (Daft, 2010).

The implication of Herzberg's two-factor theory to WLB was clear. Work–Life Enrichment could be seen as a motivator and its presence would help individuals become motivated and satisfied which eventually led to the attainment of WLB. Work–Life Conflict could be seen as a hygiene factor whereby eliminating Work–Life Conflict would help reduce individual's dissatisfaction but it does not help in attaining WLB. The application of Herzberg two-factor theory supported the results from the analysis where individuals perceived Work–Life Enrichment as an important indicator in helping them attain WLB. However, as mentioned by interviewee Miss A, there was a certain level of Work–Life Conflict limit that individual would tolerate. Once this reached the threshold, Work–Life Enrichment would not be beneficial to attaining WLB. Based on what was opined by Miss A, it could be perceived that the participants found that their level of Work–Life Conflict was tolerable and desired the presence of Work–Life Enrichment to attain Work–Life Balance to counter a tolerable level of Work–Life Conflict.

It was observed that both Singaporeans and South Koreans have different paths to achieving WLB. In the case of Singapore, the pathway included: Work–Life Enrichment → Coping Strategies → Work–Life Balance. In the case of South Korea, the pathway included: Work–Life Enrichment → Work–Life Balance. Two implications could be deduced from this result. It was felt that cultural characteristics and structure of the construction industry might be the two main differences that affected the perceptions of the Millennials employees in attaining WLB. Explanations made by interviewee Mr C have also substantiated this view when he was asked to compare the work environment between Singapore and South Korea. Mr C explained that:

> *In South Korea, all my colleagues are South Koreans and I can talk to them using my language. But in Singapore I have to use English but not all of them understand English making me very hard to talk to them especially the foreigners. There are so many foreign workers here and only a few Singapore people. My project here (Singapore) always need me to overtime until very late. When I reached home, it is already 9 pm. All my Singaporean colleagues worked until very late also. In South Korea, I will knock off at 6 pm. One thing I don't like about this project (in Singapore) is the site is very messy and dirty unlike Korea things are properly placed.*

Based on Mr C's observation, it appears that construction practices in Singapore have caused the Millennials employees difficult to attain WLB as compared to South Korea. Furthermore, another problem surfaced by Mr C was related to the huge influx of foreign workers in Singapore. Even though he only mentioned communication barriers between the foreign workers, there were many more underlying issues that affected construction productivity. These might indirectly affect the employees' productivity that could eventually affect the attainment of WLB. Cultural characteristics on the other hand could be assessed by adopting Hofstede's six cultural dimensions. For example, based on Hofstede's cultural dimension of Indulgence versus Restraint, South Korea has a score of 29 implying that it belonged to a restraint society. This explained why South Korean Millennials employees tended to have a cynicism and pessimism attitude towards their life. They were also less likely to indulge themselves due to strict social norms. As Coping Strategies in this study included behavioral management, cognitive management and leisure activities, it was more likely that South Korean Millennials employees do not find these strategies useful in achieving WLB. Further research on this issue is recommended to provide a better understanding of the reasons for the differences in attaining WLB between these two countries.

References

Aruna, M., & Anitha, J. (2015). Employee retention enablers: Generation Y employees. *SCMS Journal of Indian Management, 12*(3), 94–103.

Bae, H. J. (2017). *'No work after 6' bill to make issue in June parliament*. Retrieved June 3, 2017, from www.koreaherald.com/view.php?ud=20170531000719.

Bannon, S., Ford, K., & Meltzer, L. (2011). Understanding millennials in the workplace. *CPA Journal, 81*(11), 61–65.

Carlson, D. S., Grzywacz, J. G., & Zivnuska, S. (2009). Is work–family balance more than conflict and enrichment? *Human Relations, 62*(10), 1459–1486. https://doi.org/10.1177/0018726709336500.

Carlson, D. S., Kacmar, K. M., Wayne, J. H., & Grzywacz, J. G. (2006). Measuring the positive side of the work–family interface: Development and validation of a work–family enrichment scale. *Journal of Vocational Behavior, 68*(1), 131–164. https://doi.org/10.1016/j.jvb.2005.02.002.

Carlson, D. S., Kacmar, K., & Williams, L. J. (2000). Construction and initial validation of a multidimensional measure of work-family conflict. *Journal of Vocational Behavior, 56*(2), 249–276. https://doi.org/10.1006/jvbe.1999.1713.

Channel NewsAsia. (2016). *More companies offering flexi-work arrangements, leave benefits: MOM survey*. Retrieved June 6, 2017, from http://www.channelnewsasia.com/news/singapore/more-companies-offering-flexi-work-arrangements-leave-benefits-m-7679456.

Daft, R. (2010). *New era of management* (9th ed.). South-Western.

Frone, M. R. (2003). Work–family balance. In J. C. Quck & L. E. Tetrick (Eds.), *Handbook of occupational health psychology* (pp. 13–162). Washington, DC: American Psychological Association.

Greenhaus, J. H., & Beutell, N. J. (1985). Sources of conflict between work and family roles. *Academy of Management Review, 10*(1), 76–88. https://doi.org/10.5465/amr.1985.4277352.

Grzywacz, J. G., & Carlson, D. S. (2007). Conceptualizing work family balance: Implications for practice and research. *Advances in Developing Human Resources, 9*(4), 455–471. https://doi.org/10.1177/1523422307305487.

Haddon, B., & Hede, A. (2009). *WLB: In search of effective strategies*. Retrieved September 17, 2016 from https://www.researchgate.net/publication/272623357_Work-life_balance_In_search_of_effective_strategies.

Haddon, B., & Hede, A. (2010). *WLB—An integrated approach*. University of the Sunshine Coast.

Haddon, B., Hede, A., & Whiteoak, J. (2009). *WLB: Towards an integrated conceptual framework*. Retrieved from http://www.anzam.org/wp-content/uploads/pdf-manager/1294_HADDON_BARBARA-362.PDF.

Hair, J. F., Hult, G. T. M., Ringle, C. M., & Sarstedt, M. (2017). *A primer on partial least squares structural equation modeling* (2nd ed.). Thousand Oaks: Sage.

Health Hub. (2015). *10 easy ways to manage stress*. Retrieved June 1, 2017 from https://www.healthhub.sg/live-healthy/426/10easywaystomanagestress.

Herche, J., & Engelland, B. (1996). Reversed-polarity items and scale unidimensionality. *Journal of the Academy of Marketing Science, 24*(4), 366–374. https://doi.org/10.1177/0092070396244007.

Hofstede, G. (2001). *Culture's consequences: Comparing values, behaviors, institutions and organizations across nations*. Thousand Oaks, CA: Sage Publications.

Hofstede, G., Hofstede, J. G., & Minkov, M. (2010). *Cultures and organizations: Software of the mind* (Rev. 3 ed.). New York: McGraw-Hill.

Lee, K. C. (2014). *Factors that influence the decisions of working adults to enter the construction industry* (Unpublished undergraduate dissertation). School of Design and Environment, National University of Singapore.

Lingard, H., & Francis, V. (2009). *Managing WLB in construction*. Hoboken: Taylor and Francis.

Low, S. P. (1992). The Just-In-Time concept to improving manufacturing productivity: Is it applicable to the construction industry. *Chartered Institute of Building, 8,* 1–8.

Low, S. P., & Shi, Y. Q. (2002). An exploratory study of Hofstede's cross-cultural dimensions in construction projects. *Management Decision, 40*(1), 7–16. https://doi.org/10.1108/00251740210423036.

Lundberg, C., Gudmundson, A., & Andersson, T. D. (2009). Herzberg's Two-Factor Theory of work motivation tested empirically on seasonal workers in hospitality and tourism. *Tourism Management, 30*(6), 890–899. https://doi.org/10.1016/j.tourman.2008.12.003.

Marshall, L. L., Allison, A., Nykamp, D., & Lanke, S. (2008). Perceived stress and quality of life among doctor of pharmacy students. *American Journal of Pharmaceutical Education, 72*(6), 137. https://doi.org/10.5688/aj7206137.

Michel, J., Kotrba, L., Mitchelson, J., Clark, M., & Baltes, B. (2010). Antecedents of work-family conflict: A meta-analytic review. *Journal of Organizational Behavior, 32*(5), 689–725. https://doi.org/10.1002/job.695.

Ministry of Manpower. (2011). *Law protects employees from working excessive hour*. Retrieved June 1, 2017 from http://www.mom.gov.sg/newsroom/press-replies/2011/law-protects-employees-from-working-excessive-hour.

Olsen, P. (2008). *The world's hardest-working countries*. Retrieved June 6, 2017, from https://www.forbes.com/2008/05/21/labor-market-workforce-lead-citizen-cx_po_0521countries.html.

Opoku-Acheampong, A., Kretchy, I. A., Acheampong, F., Afrane, B. A., Ashong, S., Tamakloe, B., et al. (2017). Perceived stress and quality of life of pharmacy students in University of Ghana. *BMC Research Notes, 10*(1). https://doi.org/10.1186/s13104-017-2439-6.

Organization for Economic Co-operation and Development. (2016). *Average annual hours actually worked per worker*. Retrieved June 6, 2017, from https://stats.oecd.org/Index.aspx?DataSetCode=ANHRS.

Rantanen, J., Kinnunen, U., Mauno, S., & Tillemann, K. (2010). Introducing theoretical approaches to WLB and testing a new typology among professionals. *Creating Balance?* 27–46. https://doi.org/10.1007/978-3-642-16199-5_2.

Seah, Y. L. E. (2013). *Motivating locals to contribute to Singapore construction industry* (Unpublished undergraduate dissertation). School of Design and Environment, National University of Singapore.

Voydanoff, P. (2005a). Consequences of Boundary-Spanning demands and resources for Work-to-Family conflict and perceived stress. *Journal of Occupational Health Psychology, 10*(4), 491–503. https://doi.org/10.1037/1076-8998.10.4.491.

Voydanoff, P. (2005b). Toward a conceptualization of perceived work-family fit and balance: A demands and resources approach. *Journal of Marriage and Family, 67*(4), 822–836. https://doi.org/10.1111/j.1741-3737.2005.00178.x.

Yang, S., & Rosenblatt, P. C. (2008). Confucian family values and childless couples in South Korea. *Journal of Family Issues, 29*(5), 571–591. https://doi.org/10.1177/0192513x07309462.

Chapter 7
Conclusion

Abstract This chapter concludes by discussing the theoretical and practical implications, strengths and limitations of this study as well as recommendations for further studies. It addresses the concerns of the millennials in both Singapore and South Korea to achieve WLB in the construction industry.

7.1 Theoretical Implications

This research reviewed the relevant theories and designed a conceptual framework to operationalize the concept of WLB. The empirical studies in this research set an important way forward for the theory of WLB allowing researchers to understand the significance of each indicator associated with WLB. The results from the analysis suggested that WLE must be explored comprehensively to keep pace with the interests surrounding WLB.

Historically, WLB has been conceptualized as the absence of WLC by Grenhaus and Beutell (1985) and only just recently scholars started to explore the positive side to WLB (WLE) (Carlson et al., 2006). This study contributes to the extant literature in several ways. Firstly, the study sheds light on the interrelationships between WLC and WLE and their antecedents (demands and resources respectively). Secondly, the study provides empirical evidence of the relationships between WLC and WLE on WLB. Thirdly, the study builds on the Coping Strategies proposed by Haddon and Hede (2009) to act as a mediator between WLC and Work Life Enrichment on WLB. Finally, the study examines the relationships between WLB and Outcomes. Based on these findings, it can be seen that WLC is not significantly associated with WLB and that this differs from past research findings. Currently there is no empirical evidence to justify this reasoning and hence there is a need for more comprehensive research to establish a greater theoretical understanding of WLB. It is hoped that this study allows a shift in perspective when conceptualizing the theory of WLB. Many scholars viewed WLB as low WLC and high WLE and neglected the engagement of individual's coping strategy. Coping Strategies that incorporate environmental and individual

L. Sui Pheng and B. K. Q. Chua, *Work–Life Balance in Construction*,
Management in the Built Environment,
https://doi.org/10.1007/978-981-13-1918-1_7

characteristics are perceived to alternatively help individuals attain (perceived) WLB (Haddon & Hede, 2009).

The results of the analysis also draw another meaningful conclusion; different countries may have different ways to achieving WLB. Based on the findings, Singaporeans use Coping Strategies (via WLE) to achieving WLB while South Koreans use WLE (direct relationship) to achieving WLB. There is no one size fits all approach to assessing WLB and the results of WLB are not to be generalized. In addition, cultural characteristics associated with the targeted nationality should be assessed when evaluating WLB. The accuracy in predicting the indicators that are significant in explaining WLB has a serious implication for organizations that are implementing policies or practices within their work environment. Drawing a wrong conclusion may affect the capability of organizations in helping their employees achieve WLB and may instead result in productivity decline.

7.2 Practical Implications

The findings in this research have important implications for organizations and managers. WLB must not be seen as an overall appraisal but rather as a components approach. As mentioned earlier, an overall appraisal refers to general assessment that concerns the entirety of an individual work and life situation (Rantanen et al., 2010). An example can be “all in all, do you feel that you have achieved WLB?” Practitioners who applied this concept only attain superficial results that are too generalized to implement an effective strategy to help individuals to achieving WLB. The components approach on the other hand posits that WLB is a multi-faceted construct. This study concurs with the views of Grzywacz and Carlson (2007) as well as Rantanen et al. (2010) where WLB can be characterized as a component approach. A conceptual framework was designed in this current study to link related past theories that preceded WLB to give meaning to the concept. In so doing, this study is able to develop a comprehensive understanding in promoting WLB and to achieve valuable inputs for the organizations. It is important to note that this research focuses on Millennials Singaporeans and South Koreans working in the construction industry and the results may differ across nationalities, industries or demographics. Hence, practitioners who are applying this concept should not expect the same results but rather tweak it as a comparative study to assess the differences. The results of this analysis urge practitioners in Singapore and South Korea to focus on the positive side of WLI: WLE rather than WLC. To help individuals improve their WLE, Life and Work Resources (antecedents) should first be identified. It is hoped that the results of this study can help the practitioners to implement strategic and effective policies and practices in their organizations to promote WLB. From an organizational perspective, top management buy-in is critically essential for the success of any change effort. Top management must be supportive of the WLB initiative in order to drive changes in the policies and practices across the work environment. Some recommendations include top management constant involvement and participation in implementation, assigning a capable leader to drive the changes in the organization and empowering the leader (practitioner) to implement effective strategies to promote WLB.

7.3 Strengths and Limitations

The study has several strengths that contribute to the WLB literature in some important ways. Firstly, the relevant theories associated with WLB are reviewed to identify a conceptual framework. Even though numerous studies on WLB have been conducted, many scholars only focused on one area of WLI. As WLB is a multifaceted dimension, studying the construct partially may not yield significant outcome. Significantly, this study adds Coping Strategies into the WLB model. Coping Strategies are important to alter the effects of WLE and WLC. Although the Coping Strategies indicator is not introduced to the WLB study substantially, comments made by the interviewees showed that Coping Strategies must be present in order to help them achieve WLB especially for the Singaporean counterparts. Secondly, this study provides empirical support to justify the conceptual framework as well as to confirm observations from past studies. The use of the SmartPLS software allows all the different relationships (represented by the hypotheses) in the conceptual framework to be analyzed simultaneously. The Singapore and South Korea's results show that WLE has a stronger effect (direct or indirect) to explaining WLB. This implication allows practitioners in the construction industry to introduce strategic and effective policies and practices in their organizations that will promote WLB. Finally, the comparative study between Singapore and South Korea provides evidence that different nations may have different pathways to achieving WLB. There is no one size fits all approach to assessing WLB. The results of WLB are not generalizable to some extent. Cultural characteristics (based on Hofstede's cultural dimensions), demographics and types of industry may play a role to modifying the outcome of WLB.

No research work is without some limitations. Firstly, this study reports that Singapore has a low Uncertainty Avoidance Index (UAI) based on Hofstede's cultural dimensions. This may not be the case because it is a known fact that Singapore has various legislation, rules and regulations which are always communicated clearly to stakeholders. This should then translate to a manifestation of a high uncertainty avoidance culture and not a low UAI. In addition, Singapore is a multi-cultural society. This has some implications for the construction industry as well as adoption of Hofstede's cultural dimensions where many of the construction workers in Singapore are foreigners. Consequently, this may not translate to the respondents in Singapore having a Confucian culture to make them comparable with their counterparts in South Korea. On the other hand, it is also possible that the construction companies in Singapore practise Confucian ethos even if their workers come from various cultures.

As this study encompasses a comparative study between Singapore and South Korea, there are many difficulties in collecting data from the South Korean counterparts. English is not South Korea's first language and it is difficult for the South Koreans to understand the survey questions completely. Hence, the survey questionnaire is translated to South Korean language to ease understanding for the respondents. However, some respondents still have difficulties interpreting the questions due possibly to the incorrect vocabulary or sentence structure used. This issue may affect how respondents answer the questions. In addition, as this study provides a new perspective on the WLB research in terms of conceptual and empirical underpinnings, future research may be required to verify the results. Moreover, many areas need to be explored in order to maximize the benefit of the conceptual framework. Coping Strategies and Outcomes associated with WLB are the two indicators that can be explored further and should not be limited to just the few sub-indicators adopted in this study. Ultimately, WLB is a multifaceted dimension and the impacts or benefits associated with it are endless. Lastly, although this research only focuses on the comparison between two nations (Singapore and South Korea), the results have proved a point to future researchers and scholars that such a comparative study is possible especially in the context of international cross-border construction projects. Nevertheless, different nations, demographics of population or industries can have different ways to achieving WLB. Generalization of the result is to be avoided to some extent and practitioners must not assume that the results can be applied across other nations, demographics or industries.

7.4 Recommendations for Further Study

The WLC theory has an extensive body of research available since its first appearance in the 1980s, including prominent studies such as those of Greenhaus and Beutell (1985). However, WLE starts to gain attention only in the early 2000s, including the prominent study by Carlson et al. (2006) with empirical evidence to support the WLE model. The results of this current study show that the magnitude of effect of WLE to WLB is greater than WLC to WLB. Future research study can seek to establish an intensive study on WLE to compensate for the lack of existing literature in the WLE domain. In addition, the sub-indicators proposed in this current study were identified from past studies or formulated by the research team to suit the purpose of this study. For example, the sub-indicators for Work Resources are Job Salience, Developmental Experiences and Autonomy. The sub-indicators (or observed variables) should not be restricted to just these items. Future research studies should move beyond these sub-indicators to propose items that may be more useful in helping individuals achieve WLB. The literature review shows that WLB is an underdeveloped concept despite its widespread interests. This current study provides theoretical and empirical support to explain the concept of WLB. This study hopes that operationalizing the concept of WLB can help future researchers in this field to apply the proposed conceptual framework across other area of studies that straddle different nations, demographics or industries.

References

Carlson, D. S., Kacmar, K. M., Wayne, J. H., & Grzywacz, J. G. (2006). Measuring the positive side of the work–family interface: Development and validation of a work–family enrichment scale. *Journal of Vocational Behavior, 68*(1), 131–164. https://doi.org/10.1016/j.jvb.2005.02.002.

Greenhaus, J. H., & Beutell, N. J. (1985). Sources of conflict between work and family roles. *Academy of Management Review, 10*(1), 76–88. https://doi.org/10.5465/amr.1985.4277352.

Grzywacz, J. G., & Carlson, D. S. (2007). Conceptualizing work family balance: implications for practice and research. *Advances in Developing Human Resources, 9*(4), 455–471. https://doi.org/10.1177/1523422307305487.

Haddon, B., & Hede, A. (2009). WLB: In search of effective strategies. Retrieved September 17, 2016 from https://www.researchgate.net/publication/272623357_Work-life_balance_In_search_of_effective_strategies.

Rantanen, J., Kinnunen, U., Mauno, S., & Tillemann, K. (2010). Introducing theoretical approaches to WLB and testing a new typology among professionals. *Creating Balance?,* 27–46. https://doi.org/10.1007/978-3-642-16199-5_2.

Appendix A
Average Weekly Total Paid Hours Worked Per Employee by Industry

L. Sui Pheng and B. K. Q. Chua, *Work–Life Balance in Construction*,
Management in the Built Environment,
https://doi.org/10.1007/978-981-13-1918-1

AVERAGE WEEKLY TOTAL PAID HOURS WORKED PER EMPLOYEE BY INDUSTRY

Hours

Industry (SSIC 2010)		2013	2014	2015	2014	2015			
					Dec	Mar	Jun	Sep	Dec
	TOTAL	**46.2**	**46.0**	**45.6**	**45.8**	**45.6**	**45.6**	**45.6**	**45.5**
C10-32	**MANUFACTURING**	**50.1**	**49.7**	**49.3**	**49.7**	**49.3**	**49.4**	**49.3**	**49.0**
C10-12	Food, Beverages & Tobacco	47.0	47.0	46.6	47.3	46.4	46.7	47.1	46.3
C17,18,22	Paper / Rubber / Plastic Products & Printing	49.1	49.3	48.5	48.9	48.7	48.4	48.7	48.3
C19-21	Petroleum, Chemical & Pharmaceutical Products	44.7	44.5	44.4	44.4	44.3	44.3	44.4	44.5
C25,28	Fabricated Metal Products, Machinery & Equipment	52.6	52.1	51.0	51.8	50.9	51.2	51.1	50.8
C26	Electronic, Computer & Optical Products	46.9	46.8	46.4	46.5	46.4	46.4	46.5	46.3
C29-30	Transport Equipment	53.1	52.4	52.7	52.9	53.0	52.7	52.6	52.4
	Other Manufacturing Industries	50.2	49.2	48.5	49.1	48.3	49.3	48.3	48.3
F41-43	**CONSTRUCTION**	**53.2**	**52.9**	**52.1**	**52.4**	**52.0**	**52.2**	**52.1**	**51.9**
G-U	**SERVICES**	**43.4**	**43.2**	**43.1**	**43.1**	**43.1**	**43.0**	**43.1**	**43.1**
G46-47	WHOLESALE AND RETAIL TRADE	43.2	43.1	43.0	43.1	43.0	43.3	43.1	42.6
G46	Wholesale Trade	43.6	43.6	43.4	43.8	43.5	43.7	43.5	43.0
G47	Retail Trade	42.3	42.0	42.1	41.6	41.8	42.4	42.3	41.8
H49-53	TRANSPORTATION AND STORAGE	45.7	45.8	45.7	45.9	45.8	45.3	45.7	46.1
H49,5221	Land Transport & Supporting Services	46.8	46.9	46.0	46.6	46.3	45.3	46.2	46.2
H50,5222, 5225	Water Transport & Supporting Services	44.3	44.5	44.3	44.9	44.7	44.2	44.0	44.5
H51,5223	Air Transport & Supporting Services	45.1	45.0	45.2	44.7	45.6	44.4	44.7	46.0
	Other Transportation & Storage Services	46.7	46.7	47.1	46.9	46.6	46.8	47.6	47.4
I55-56	ACCOMMODATION AND FOOD SERVICES	43.0	42.0	41.5	42.0	41.4	41.3	41.5	41.9
I55	Accommodation	46.6	46.5	46.4	46.3	46.5	46.4	46.5	46.4
I56	Food & Beverage Services	42.0	40.9	40.4	41.1	40.2	40.1	40.4	40.9
J58-63	INFORMATION AND COMMUNICATIONS	41.7	41.5	41.5	41.3	41.4	41.5	41.7	41.5
J58-61	Telecommunications, Broadcasting & Publishing	42.3	41.9	41.9	41.9	41.9	42.0	42.1	41.8
J62-63	IT & Other Information Services	41.2	41.2	41.2	41.0	41.0	41.1	41.4	41.3
K64-66	FINANCIAL AND INSURANCE SERVICES	41.2	41.1	41.1	41.1	41.1	41.1	41.0	41.1
K64 & 66 (excl.662)	Financial Services	41.3	41.3	41.2	41.2	41.2	41.3	41.2	41.3
K65 & 662	Insurance Services	40.0	40.2	40.1	40.1	40.2	40.1	40.0	40.2
L68	REAL ESTATE SERVICES	44.6	44.5	44.8	44.4	44.6	45.1	44.8	44.9
M69-75	PROFESSIONAL SERVICES	43.6	43.1	42.9	43.0	43.0	42.7	42.9	43.1
M69-70	Legal, Accounting & Management Services	41.3	41.2	41.2	41.5	41.5	41.1	41.2	41.2
M71	Architectural & Engineering Services	47.0	46.2	45.6	45.8	45.5	45.4	45.5	46.0
	Other Professional Services	41.8	41.7	42.0	41.6	41.6	41.8	42.4	42.3
N77-82	ADMINISTRATIVE AND SUPPORT SERVICES	47.5	47.3	46.4	46.7	46.6	46.3	46.3	46.4
N80	Security & Investigation	52.8	52.1	50.5	49.8	51.1	50.2	50.4	50.4
N81	Cleaning & Landscaping	44.9	44.6	43.8	45.0	43.9	43.6	43.8	44.0
	Other Administrative & Support Services	44.8	45.2	45.3	45.5	45.2	45.3	45.4	45.3
O-U	COMMUNITY, SOCIAL AND PERSONAL SERVICES	42.1	42.0	42.1	42.0	42.1	42.1	42.1	42.1
O84,P85	Public Administration & Education	41.3	41.3	41.5	41.3	41.4	41.5	41.5	41.6
Q86-88	Health & Social Services	42.0	41.9	41.9	41.9	41.9	41.9	41.9	41.9
R90-93	Arts, Entertainment & Recreation	43.9	43.8	44.2	44.0	44.2	44.2	44.2	44.1
S,T,U	Other Community, Social & Personal Services	43.8	43.5	42.9	43.7	43.3	43.0	42.9	42.6
A,B,D,E,V	**OTHERS***	**45.7**	**45.6**	**45.6**	**45.8**	**45.6**	**45.6**	**45.4**	**45.8**

(Source: Ministry of Manpower, 2016b)

Appendix B
Average Weekly Paid Overtime Hours Worked Per Employee by Industry

L. Sui Pheng and B. K. Q. Chua, *Work–Life Balance in Construction*,
Management in the Built Environment,
https://doi.org/10.1007/978-981-13-1918-1

AVERAGE WEEKLY PAID OVERTIME HOURS WORKED PER EMPLOYEE BY INDUSTRY

Hours

Industry (SSIC 2010)		2013	2014	2015	2014	2015			
					Dec	Mar	Jun	Sep	Dec
	TOTAL	3.8	3.7	3.4	3.6	3.4	3.4	3.4	3.4
C10-32	**MANUFACTURING**	**7.1**	**6.7**	**6.3**	**6.7**	**6.4**	**6.4**	**6.4**	**6.1**
C10-12	Food, Beverages & Tobacco	4.1	3.9	4.0	4.2	3.6	4.0	4.3	3.9
C17,18,22	Paper / Rubber / Plastic Products & Printing	5.7	5.8	5.4	5.4	5.6	5.3	5.6	5.1
C19-21	Petroleum, Chemical & Pharmaceutical Products	2.9	2.8	2.6	2.8	2.6	2.6	2.5	2.7
C25,28	Fabricated Metal Products, Machinery & Equipment	9.0	8.6	7.6	8.3	7.5	7.8	7.7	7.3
C26	Electronic, Computer & Optical Products	4.8	4.8	4.3	4.5	4.5	4.4	4.4	4.0
C29-30	Transport Equipment	9.5	8.8	9.0	9.1	9.2	8.9	9.0	8.9
	Other Manufacturing Industries	7.2	6.3	5.9	6.2	5.7	6.5	5.8	5.8
F41-43	**CONSTRUCTION**	**8.5**	**8.2**	**7.6**	**7.8**	**7.3**	**7.5**	**7.7**	**7.7**
G-U	**SERVICES**	**1.7**	**1.7**	**1.6**	**1.7**	**1.6**	**1.6**	**1.6**	**1.6**
G46-47	WHOLESALE AND RETAIL TRADE	1.7	1.8	1.8	2.1	1.8	2.0	1.9	1.6
G46	Wholesale Trade	1.5	1.7	1.8	2.1	1.8	2.0	1.9	1.5
G47	Retail Trade	2.0	2.0	1.8	2.0	1.8	1.9	1.8	1.8
H49-53	TRANSPORTATION AND STORAGE	3.0	3.0	2.9	2.9	2.9	2.7	2.8	3.2
H49,5221	Land Transport & Supporting Services	3.8	4.0	3.3	3.5	3.3	3.3	3.3	3.4
H50,5222, 5225	Water Transport & Supporting Services	1.7	1.9	1.7	2.0	1.8	1.7	1.7	1.8
H51,5223	Air Transport & Supporting Services	2.2	1.8	2.1	1.4	2.4	1.3	1.8	2.9
	Other Transportation & Storage Services	4.0	4.2	4.1	4.1	3.8	4.1	4.2	4.3
I55-56	ACCOMMODATION AND FOOD SERVICES	2.3	1.9	1.8	1.8	1.8	1.8	1.7	1.9
I55	Accommodation	2.6	2.7	2.6	2.6	2.7	2.5	2.6	2.7
I56	Food & Beverage Services	2.2	1.7	1.6	1.6	1.6	1.6	1.5	1.7
J58-63	INFORMATION AND COMMUNICATIONS	0.5	0.4	0.4	0.3	0.4	0.4	0.4	0.4
J58-61	Telecommunications, Broadcasting & Publishing	0.8	0.6	0.7	0.4	0.6	0.7	0.7	0.6
J62-63	IT & Other Information Services	0.3	0.2	0.2	0.1	0.2	0.2	0.2	0.2
K64-66	FINANCIAL AND INSURANCE SERVICES	0.2	0.2	0.2	0.2	0.2	0.2	0.2	0.2
K64 & 66 (excl.662)	Financial Services	0.2	0.2	0.2	0.2	0.2	0.2	0.2	0.2
K65 & 662	Insurance Services	0.3	0.3	0.2	0.3	0.3	0.2	0.2	0.2
L68	REAL ESTATE SERVICES	2.2	2.0	2.3	2.0	2.2	2.5	2.2	2.3
M69-75	PROFESSIONAL SERVICES	2.1	1.8	1.4	1.6	1.5	1.3	1.3	1.5
M69-70	Legal, Accounting & Management Services	0.8	0.8	0.5	0.6	0.7	0.5	0.5	0.6
M71	Architectural & Engineering Services	4.3	3.7	2.9	3.2	2.9	2.7	2.7	3.2
	Other Professional Services	0.7	0.6	0.7	0.7	0.7	0.8	0.8	0.8
N77-82	ADMINISTRATIVE AND SUPPORT SERVICES	4.8	5.3	4.4	5.1	4.7	4.5	4.2	4.4
N80	Security & Investigation	9.6	10.3	9.1	9.4	10.0	9.1	8.6	8.6
N81	Cleaning & Landscaping	2.3	2.6	1.8	3.1	1.8	1.7	1.6	2.2
	Other Administrative & Support Services	2.6	2.9	2.8	3.1	2.6	2.9	2.9	2.9
O-U	COMMUNITY, SOCIAL AND PERSONAL SERVICES	0.5	0.5	0.5	0.5	0.5	0.5	0.5	0.5
O84,P85	Public Administration & Education	0.1	0.1	0.1	0.1	0.1	0.1	0.1	0.1
Q86-88	Health & Social Services	0.4	0.5	0.5	0.5	0.5	0.5	0.5	0.4
R90-93	Arts, Entertainment & Recreation	1.2	1.2	1.5	1.4	1.5	1.4	1.3	1.7
S,T,U	Other Community, Social & Personal Services	1.7	1.7	1.5	2.0	1.8	1.7	1.8	1.0
A,B,D,E,V	**OTHERS***	**3.1**	**2.9**	**3.0**	**3.0**	**3.0**	**3.0**	**2.7**	**3.2**

(Source: Ministry of Manpower, 2016b)

Appendix C
Age Pyramid of the Resident Population in 2005 and 2015

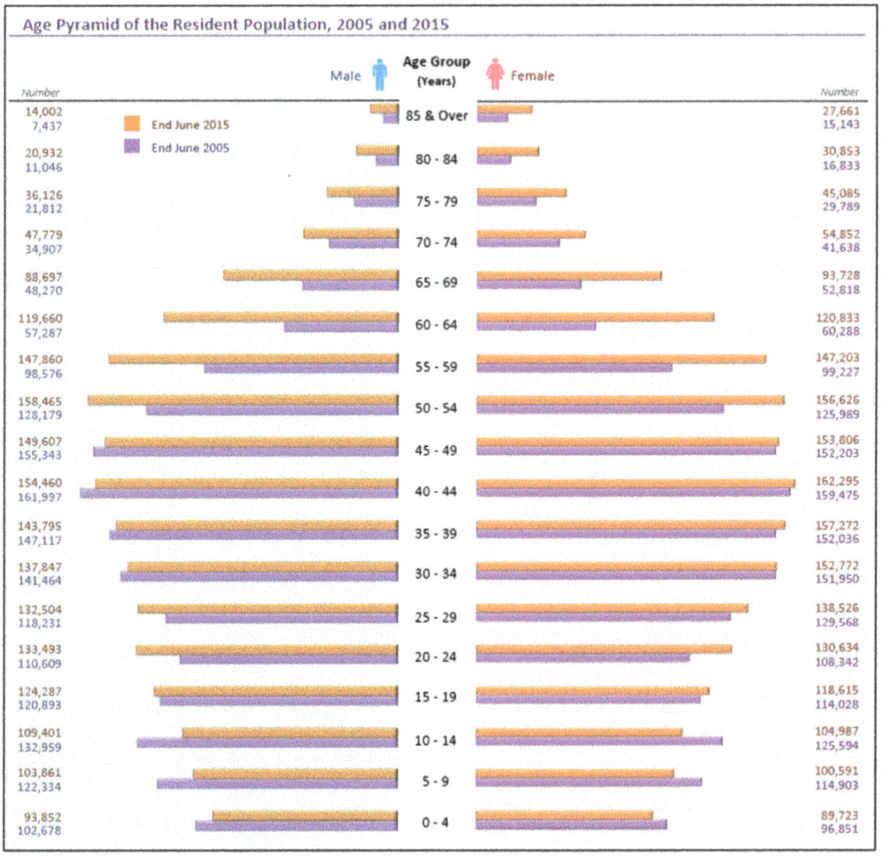

(Source: Department of Statistics Singapore, 2016)

L. Sui Pheng and B. K. Q. Chua, *Work–Life Balance in Construction*,
Management in the Built Environment,
https://doi.org/10.1007/978-981-13-1918-1

Appendix D
Survey on Wage and Working Hours by Industry

	Hours worked of permanent employees per month	Regular worked of permanent employees (Hours)	Overtime hours Worked of permanent employees (Hours)	Total wages of permanent employees (Won)	Regular wages of permanent employees (Won)	Overtime wages of permanent employees (Won)
Total	187.9	174.9	13.0	2,926,186	2,710,026	216,160
Mining and quarrying	192.0	174.1	17.9	3,138,506	2,847,871	290,635
Manufacturing	197.3	172.2	25.1	2,938,589	2,536,392	402,197
Electricity gas steam and water supply	182.9	171.5	11.4	4,562,234	4,253,706	308,528
Sewerage waste management materials recovery activities	187.8	175.6	12.2	2,727,016	2,492,337	234,679
Construction	181.9	177.3	4.7	3,011,771	2,913,538	98,232
Wholesale and retail trade	184.9	178.9	6.0	2,908,227	2,811,228	96,999
Transportation	180.8	167.9	12.9	2,534,964	2,285,733	249,231
Accommodation and food service activities	205.9	199.1	6.8	2,132,191	2,033,963	98,228
Information and communications	178.0	172.3	5.7	3,644,677	3,507,530	137,147
Financial and insurance activities	175.6	170.1	5.5	3,894,984	3,772,783	122,201
Real estate activities and renting and leasing	199.2	196.8	2.4	2,329,398	2,271,250	58,148
Professional scientific and technical activities	175.5	171.7	3.8	3,955,201	3,889,244	65,957
Business facilities management and support services	186.7	176.6	10.0	1,905,393	1,776,906	128,487
Education	180.2	172.9	7.2	3,389,946	3,305,485	84,461
Human health and social work activities	181.9	174.7	7.2	2,532,599	2,368,698	163,901
Arts sports and recreation related services	187.7	174.3	13.4	2,637,147	2,436,318	200,829
Other personal services	186.4	181.5	4.9	2,300,340	2,218,634	81,706

(Source: Ministry of Employment and Labour, n.d.)

L. Sui Pheng and B. K. Q. Chua, *Work–Life Balance in Construction*,
Management in the Built Environment,
https://doi.org/10.1007/978-981-13-1918-1

Appendix E
Survey Questionnaire in English Version

L. Sui Pheng and B. K. Q. Chua, *Work–Life Balance in Construction*,
Management in the Built Environment,
https://doi.org/10.1007/978-981-13-1918-1

NATIONAL UNIVERSITY OF SINGAPORE
School of Design and Environment
Department of Building

A STUDY ON WORK-LIFE BALANCE IN SOUTH KOREA AND SINGAPORE

This study aims to find out the level of Work-Life Balance among the Generation Yís Construction Professionals in South Korea and Singapore. Series of questions will be asked to capture each respondentís perception of their work and life domain. After the comprehensive survey, a comparative study will be made to examine the level of Work-Life Balance between the two nations.

Definition:

- **Work-Life Balance**: A comfortable state of equilibrium between Work and Life domain achieved by individual. This means that demand in work must not be overwhelmed by individualís ability to satisfy his/her personal life outside the work environment or vice versa.
- **Generation Y**: A group of population who are born after 1980s.

All responses gathered are highly regarded and will be analyzed thoroughly to ensure the accuracy of this research. **All data will be kept strictly confidential and will only be used for academic research purposes**. Should you have any queries, please contact Benjamin Chua at 9026XXXX or via email at Benjamin_Chua@u.nus.edu.

SECTION A — PERSONAL PARTICULARS

Name of Respondent : ______________________________

Name of Company : ______________ **Age:** : ______________

Profession : ______________ **Email** : ______________

SECTION B — WORK LIFE BALANCE												
	Part A — Work-Domain	=1 Strongly Disagree	=2 Disagree	=3 Neither Agree nor Disagree	=4 Agree	=5 Strongly Agree	Part B — Life-Domain	=1 Strongly Disagree	=2 Disagree	=3 Neither Agree nor Disagree	=4 Agree	=5 Strongly Agree
a)	**Work Demands**						**Life Demands**					
	Work Role Conflict						**Life Role Conflict**					
1	I do things that are apt to be accepted by one person and not accepted by others	1	2	3	4	5	I do things that are apt to be accepted by one person and not accepted by others	1	2	3	4	5
2	I work under incompatible policies and guidelines	1	2	3	4	5	I have many roles in life that operate quite differently from one another	1	2	3	4	5
	Work Role Ambiguity						**Life Role Ambiguity**					
3	I do not know exactly what my responsibilities are at work	1	2	3	4	5	I do not know exactly what my responsibilities are in life	1	2	3	4	5
4	I am uncertain as to how my job is linked	1	2	3	4	5	I do not know exactly what is expected of me in life	1	2	3	4	5
	Work Involvement						**Life Involvement**					
5	I would like to spend more time working	1	2	3	4	5	I would not like to spend more time in my non-work life	1	2	3	4	5
6	I have many roles in work that operate quite differently from one another	1	2	3	4	5	The most important things that happen to me involve my life	1	2	3	4	5
b)	**Boundary Spanning Demands**											
	Work Overload						**Life Activities At Work**					
7	My job has often overload me with works that must be completed within a tight deadline	1	2	3	4	5	My job allows me to bring life activities to work (eg. Communicating with friends and family on phone)	1	2	3	4	5
8	Overloading with work in my job has negatively affected my personal life	1	2	3	4	5	Bringing life activities to work has negatively affected my working life	1	2	3	4	5
	Bring Work to Home						**Life Interruptions and Distraction**					
9	I often bring work to home as the project needs me to be available 24/7	1	2	3	4	5	I often allow my personal life issues to be brought over to work	1	2	3	4	5

10	Bringing work to home has negatively affected my personal life	1	2	3	4	5	I often get interrupted or distracted by my personal life issues at work	1	2	3	4	5
c)	**Work Resources**						**Life Resources**					
	Job Salience						**Life Salience**					
11	The major satisfaction in my life comes from my job	1	2	3	4	5	Most things in my life are more important than work	1	2	3	4	5
12	I am very much involved personally in my work	1	2	3	4	5	I am very much involved personally in my life	1	2	3	4	5
	Developmental Experiences						**Life Harmony**					
13	At the position that I have held at this organization, I have often been assigned projects that enabled me to develop and strengthen new skills	1	2	3	4	5	I feel that I am living in a harmonious life	1	2	3	4	5
14	At the position that I have held at this organization, I have often been given additional challenging assignments.	1	2	3	4	5	My duty as a friend, family member or other non-work role is co-operating with each other harmoniously	1	2	3	4	5
	Autonomy						**Life (Psychological) Rewards**					
15	I have significant autonomy in determining how I do my job	1	2	3	4	5	I am satisfied with the recognition that I gotten from my families and friends	1	2	3	4	5
16	I have considerable opportunity for independence and freedom in how I do my job	1	2	3	4	5	I am satisfied with the compliments from my families and friends	1	2	3	4	5
d)	**Boundary Spanning Resources**											
	Flexible Work Schedule						**Participate in Workshops for Personal Development**					
17	I am given flexible work schedule whereby I can work at a time that best suits me	1	2	3	4	5	I often participate in workshops for personal development	1	2	3	4	5
18	Flexible work schedule has positively affected my personal life	1	2	3	4	5	Participating in workshops for personal development has positively affected my working and personal life	1	2	3	4	5
	Time-off for Non-Work Purposes						**Assigned Time for Work-Purposes**					
19	My organization allowed me to apply time-off for non-work purposes	1	2	3	4	5	I will assign time during my free days for work purposes	1	2	3	4	5
20	Having time-off for non-work purposes has positively affected personal life	1	2	3	4	5	Assigning time for work purposes have positively affected my working and personal life	1	2	3	4	5

e)	**Proposed Measures** ▪ **Reduce Demands**											
	Take Less Demanding Task						**Reduce Time With Love Ones**					
21	I prefer taking up less demanding task than the work I am dealing right now	1	2	3	4	5	Reduce time with my love ones will allow me to better focus on my work	1	2	3	4	5
	No Work After 6						**Limit Usual Lifestyles**					
22	I hope to shut off from work after 6pm	1	2	3	4	5	Limit my usual lifestyle will allow me to have better focus on my work	1	2	3	4	5
	Organization Plays A Part						**Reduce Role Responsibility**					
23	I hope my organization can play a bigger role in helping me cope with the high demand at work	1	2	3	4	5	Reduce responsibility in my non-work life will help me attain balance between my work and life	1	2	3	4	5
	Proposed Measures ▪ **Increase Resources**											
	Assign More Challenging Task						**Take Up New Sports/Hobby**					
24	I want to have more challenging tasks assign to me	1	2	3	4	5	I wish to take up new sports and hobbies to switch my mind off work	1	2	3	4	5
	More Break Interval						**Planning my weekly activities**					
25	I hope to have more breaks in between my working hour to allow me to relax my mind	1	2	3	4	5	Planning weekly activities with friends and family will allow me to have something to look forward to	1	2	3	4	5
	Provide More Freedom						**Assign More Personal Time**					
26	Having more freedom in my work allows me to produce better result	1	2	3	4	5	I want to have more uninterrupted personal time during my free time to keep away from work and life	1	2	3	4	5
f)	**Work-Life Conflict**											
	Time-based Work to Life Conflict						**Time-based Life to Work Conflict**					
27	My work keeps me from my life activities more than I would like	1	2	3	4	5	The time I spend on life responsibilities often interfere with my work responsibilities	1	2	3	4	5
28	The time I must devote to my job keeps me from participating equally in household responsibilities and activities	1	2	3	4	5	The time I spend with my life often causes me not to spend time in activities at work that could be helpful to my career	1	2	3	4	5
29	I have to miss life activities due to the amount of time I must spend on work responsibilities	1	2	3	4	5	I have to miss work activities due to the amount of time I must spend on life responsibilities	1	2	3	4	5

	Strain-based Work to Life Conflict						**Strain-based Life to Work Conflict**					
30	When I get home from work I am often too frazzled to participate in life activities/responsibilities	1	2	3	4	5	Due to stress at home, I am often preoccupied with life matters at work	1	2	3	4	5
31	I am often so emotionally drained when I get home from work that it prevents me from contributing to my life	1	2	3	4	5	Because I am often stressed from life responsibilities, I have a hard time concentrating on my work	1	2	3	4	5
32	Due to all the pressures at work, sometimes when I come home I am too stressed to do the things I enjoy	1	2	3	4	5	Tension and anxiety from my life often weakens my ability to do my job	1	2	3	4	5
	Behavior-based Work to Life Conflict						**Behavior-based Life to Work Conflict**					
33	The problem-solving behaviors I use in my job are not effective in resolving problems at home	1	2	3	4	5	The behaviors that work for me at home do not seem to be effective at work	1	2	3	4	5
34	Behavior that is effective and necessary for me at work would be counterproductive at home	1	2	3	4	5	Behavior that is effective and necessary for me at home would be counterproductive at work	1	2	3	4	5
35	The behaviors I perform that make me effective at work do not help me to be a better person in my non-work life	1	2	3	4	5	The problem-solving behavior that works for me at home does not seem to be as useful at work	1	2	3	4	5
g)	**Work Life Enrichment**											
	***Question: My involvement in my work / life ____________________.**											
	Work to Life Development						**Life to Work Development**					
36	Helps me to understand different viewpoints and this helps me be a better person in my life	1	2	3	4	5	Helps me to gain knowledge and this helps me be a better worker	1	2	3	4	5
37	Helps me to gain knowledge and this helps me be a better person in my life	1	2	3	4	5	Helps me acquire skills and this helps me be a better worker	1	2	3	4	5
38	Helps me acquire skills and this helps me be a better person in my life	1	2	3	4	5	Helps me expand my knowledge of new things and this helps me be a better worker	1	2	3	4	5
	Work to Life Affect						**Life to Work Affect**					
39	Puts me in a good mood and this helps me be a better person in my life	1	2	3	4	5	Puts me in a good mood and this helps me be a better worker	1	2	3	4	5
40	Makes me feel happy and this helps me be a better person in my life	1	2	3	4	5	Makes me feel happy and this helps me be a better worker	1	2	3	4	5

41	Makes me cheerful and this helps me be a better person in my life	1	2	3	4	5	Makes me cheerful and this helps me be a better worker	1	2	3	4	5
	Work to Life Capital						**Life to Work Efficiency**					
42	Helps me feel personally fulfilled and this helps me be a better person in my life	1	2	3	4	5	Requires me to avoid wasting time at work and this helps me be a better worker	1	2	3	4	5
43	Provides me with a sense of accomplishment and this helps me be a better person in my life	1	2	3	4	5	Encourages me to use my work time in a focused manner and this helps me be a better worker	1	2	3	4	5
44	Provides me with a sense of success and this helps me be a better person in my life	1	2	3	4	5	Causes me to be more focused at work and this helps me be a better worker	1	2	3	4	5

Part C – Overall Assessment of Work-Life Balance												
h)	**Coping Strategies**											
	Colleague and Employer Support						**Families and Friends Support**					
45	Support from my colleagues and employer have helped me better cope with stress and improve my performance in work and life roles	1	2	3	4	5	Support from my friends and families have helped me better cope with stress and improve my performance in work and life roles	1	2	3	4	5
	Leisure Activities						**Cognitive Management**					
46	My usual leisure activities have helped me better cope with stress and improve my performance in work and life roles	1	2	3	4	5	I am used to the usual work and life habits that make me have the ability to deal with high stress or low performance at work or life	1	2	3	4	5
	Behavioral Management											
47	I will engage in specific behavior which is appropriate to deal with high stress or low performance at work or life	1	2	3	4	5						

i)	**Work Life Balance**											
48	I am able to negotiate and accomplish what is expected of me at work and in my life	1	2	3	4	5	I am able to accomplish the expectations that my superior and my family have for me	1	2	3	4	5
49	I do a good job of meeting the role expectations of critical people in my work and life	1	2	3	4	5	My co-workers and friends and family would say that I am meeting their expectations	1	2	3	4	5
50	People who are close to me would say that I do a good job of balancing work and life	1	2	3	4	5	It is clear to me, based on feedback from co-workers, friends and family members, that I am accomplishing both my work and life responsibilities	1	2	3	4	5

j)	**Work-Domain Outcomes**						**Life-Domain Outcomes**					
	Job Satisfaction						**Family Satisfaction**					
51	The work I do on my job is meaningful to me	1	2	3	4	5	The conditions of my family life are excellent	1	2	3	4	5
52	All in all, I am satisfied with my job	1	2	3	4	5	I am satisfied with my family life	1	2	3	4	5
	Organizational Commitment						**Life Satisfaction**					
53	I am proud to be able to tell people who it is I work for	1	2	3	4	5	I am satisfied with my life	1	2	3	4	5
54	I really feel as if my organization's problem is my own.	1	2	3	4	5	In most ways my life is close to my ideal	1	2	3	4	5
	Intention to Turnover						**Psychological Well-Being**					
55	It is likely that I will be actively looking for new job next year	1	2	3	4	5	In general, I feel confident and positive about myself	1	2	3	4	5
56	I will not give up on this job easily	1	2	3	4	5	For the most part, I am proud of whom I am and the life I lead	1	2	3	4	5

☐ I would like to receive a summary report of this study and please send a copy to my above email address

Thank you very much for completing this questionnaire.

Appendix F
Survey Questionnaire in Korean Version

L. Sui Pheng and B. K. Q. Chua, *Work–Life Balance in Construction*,
Management in the Built Environment,
https://doi.org/10.1007/978-981-13-1918-1

NATIONAL UNIVERSITY OF SINGAPORE
School of Design and Environment
Department of Building

한국과 싱가포르의 일과 삶의 균형에 관한 연구

본 연구는 한국과 싱가포르의 Y 세대 건축 전문가들 사이의 일과 삶의 균형을 찾아내는 것을 목표로 합니다. 앞으로의 질문은 응답자의 일과 삶의 영역에 대한 인식을 수집하기 위하여 요청 될 것입니다. 종합 조사 후, 양국의 일과 삶의 균형을 조사하기 위한 비교 연구가 실시 될 것 입니다.

정의 :

- **일과 삶의 균형 (WLB) : 개인이 성취 한 일과 삶의 영역 사이에서 균형이 잡힌 편안한 상태. 즉, 직장에서의 요구가 직장 환경 밖에서 개인의 삶을 만족시킬 수있는 능력에 의해 압도되어서는 안되며, 그 반대도 마찬가지입니다.**
- **Y 세대 : 1980 년대 이후에 태어난 인구 집단.**

수집 된 모든 응답은 높은 평가를 받고이 연구의 정확성을 보장하기 위해 철저히 분석 될 것입니다. 모든 데이터는 엄격하게 기밀로 유지되며 학술 연구 목적으로만 사용됩니다. 문의 사항이 있으시면 Benjamin Chua에게 010-2769-XXXX 로 전화 하시거나 이메일 Benjamin_Chua@u.nus.edu로 연락주세요.

구간 A – 개인 세부사항

응답자 이름 : ______________________

회사 이름 : ______________ **나이** : ______________

직업 : ______________ **이메일** : ______________

구간 B – 일과 삶의 균형												
	부분 A – 일의 영역	=1 매우 동의하지 않는다	=2 동의하지 않는다	=3 보통이다	=4 동의한다	=5 아주 동의한다	부분 B – 삶의 영역	=1 매우 동의하지 않는다	=2 동의하지 않는다	=3 보통이다	=4 동의한다	=5 아주 동의한다
a)	**일의 요구들**						**삶의 요구들**					
	일의 역할에서 갈등						**삶의 역할에서 갈등**					
1	나는 누군가 에게는 받아 들여지기 쉽고 다른 사람들에게는 받아 들여지지 않는 일을 한다.	1	2	3	4	5	나는 누군가 에게는 받아 들여지기 쉽고 다른 사람들에게는 받아 들여지지 않는 일을 한다.	1	2	3	4	5
2	나는 양립 할 수 없는 정책과 지침에 따라 일한다.	1	2	3	4	5	나는 서로 다르게 운영되는 삶에서 많은 역할을 맡고 있다.	1	2	3	4	5
	일 역할의 모호성						**삶 역할의 모호성**					
3	나는 내 책임이 무엇인지 정확히 모른다.	1	2	3	4	5	나는 내 책임이 무엇인지 정확히 모른다.	1	2	3	4	5
4	내 직업이 어떻게 연관되어있는지 확신할 수 없다.	1	2	3	4	5	나는 나에게 기대되는 것이 정확히 무엇인지 모른다.	1	2	3	4	5
	일의 개입						**삶의 개입**					
5	나는 일하는 데 더 많은 시간을 보내고 싶다.	1	2	3	4	5	나는 비 직장 생활에서 더 많은 시간을 보내고 싶지 않을 것이다.	1	2	3	4	5
6	나는 서로 다르게 운영되는 삶에서 많은 역할을 맡고 있다.	1	2	3	4	5	내게 일어나는 가장 중요한 일은 내 삶과 관련이있다.	1	2	3	4	5
b)	**경계를 넘나드는 요구**											
	업무과다						**직장에서의 일상생활들**					
7	나의 일은 종종 마감 시간 내에 완료해야 하는 작업으로 인해	1	2	3	4	5	직장에서 일상생활들을 할 수 있다 (예 : 친구 및 가족과 전화 통화)	1	2	3	4	5

	업무과다 상태가 되었다.											
8	나의 업무에 있어서 업무 과다가 생긴 경우 개인적인 삶에 부정적인 영향을 미쳤다.	1	2	3	4	5	직장에서 일상생활들을 하는 것은 직장생활에 부정적인 영향을 미쳤다.	1	2	3	4	5
	집에서의 업무						**삶의 방해 및 산만**					
9	프로젝트가 날 상시로 필요로 하기 때문에 종종 집에서 업무를 처리하곤 한다.	1	2	3	4	5	나는 종종 내 개인적인 삶의 문제를 일터로 가져 가게 한다.	1	2	3	4	5
10	집에서 업무를 처리하는 것 은 내 개인적인 생활에 부정적인 영향을 미치고 있다.	1	2	3	4	5	나는 종종 직장에서의 개인적인 삶의 문제로 인해 방해 받거나 혼란스러워진다.	1	2	3	4	5
c)	**일의 자원**						**삶의 자원**					
	직업 부각성						**삶의 부각성**					
11	나는 직장에서 사생활 문제로 인해 종종 방해 받거나 혼란스러워 진다.	1	2	3	4	5	내 인생에서 대부분의 것 들은 일보다 더 중요하다.	1	2	3	4	5
12	나는 내 일에 개인적으로 매우 관여한다.	1	2	3	4	5	나는 내 인생에 개인적으로 매우 관여한다.	1	2	3	4	5
	개발적인 경험						**삶의 조화**					
13	내가 이 조직에서 맡고 있는 직책에서 나는 새로운 기술을 개발하고 강화할 수 있는 프로젝트를 종종 배정 받고 있다.	1	2	3	4	5	나는 조화로운 삶을 살고 있다고 느낀다.	1	2	3	4	5
14	내가 이 조직에서 맡고 있는 직책에서 나는 더욱 더 도전적인 과제를 부여 받고 있다.	1	2	3	4	5	친구, 가족 또는 다른 역할의 구성원으로서 나의 의무는 서로 조화롭게 협력하고 있다.	1	2	3	4	5
	자율권						**생활 (심리적) 보상**					
15	나는 내 업무방식을 결정하는 데	1	2	3	4	5	나는 가족과 친구들에게서 받은	1	2	3	4	5

	있어서 큰 자율성을 가지고 있다.						인정에 만족한다.					
16	나는 나의 일하는 방식에 있어서 독립과 자유를 누릴 수 있는 상당한 기회를 가지고 있다.	1	2	3	4	5	나는 가족과 친구들의 칭찬에 만족한다.	1	2	3	4	5
d)	**경계를 분리하는 자원들**											
	유연한 근무 일정						**자기개발을 위한 연수회 참여**					
17	나는 내가 가장 잘 맞는 시간에 일할 수 있는 유연한 근무 일정을 받는다.	1	2	3	4	5	나는 종종 자기개발을 위해 연수회에 참여한다.	1	2	3	4	5
18	유연한 근무 일정은 내 개인적인 삶에 긍정적 인 영향을 미치고 있다.	1	2	3	4	5	자기개발을 위해 연수회에 참여하는 것이 내 일과 개인 생활에 긍정적 인 영향을 미쳤다.	1	2	3	4	5
	비 업무용 휴식						**업무를 위해 할당된 시간**					
19	나의 조직은 비 업무적인 일을 위해 휴직을 신청할 수 있다.	1	2	3	4	5	업무를 위하여 쉬는 날에도 시간을 할애하겠다.	1	2	3	4	5
20	비 업무를 위해 시간을 내는 것은 개인 생활에 긍정적 인 영향을 미치고 있다.	1	2	3	4	5	업무를 목적으로 시간을 할당하는 것 은 일과 개인 생활에 긍정적 인 영향을 미치고 있다.	1	2	3	4	5
e)	**제안된 대책** ▪ **요구를 감소**											
	덜 까다로운 작업 가져 가기						**애정으로 시간 단축**					
21	나는 지금 다루고있는 일보다 덜 까다로운 일을 선호한다.	1	2	3	4	5	내 애정으로 시간을 단축하면 내 작업에 더 집중할 수 있다.	1	2	3	4	5
	6시 이후 무근						**일상생활 제한**					
22	오후 6시 이후에 업무를 중단하길 원한다.	1	2	3	4	5	내 일상 생활을 제한하면 내 작업에 더 집중할 수 있다.	1	2	3	4	5

	A역할을 하는 조직						**역할책임의 감소**					
23	일의 높은 수요에 대처할 때 내 조직이 더 큰 역할을 할 수 있기를 바란다.	1	2	3	4	5	비 직장 생활에서 책임을 줄이면 일과 삶의 균형을 이룰 수 있다.	1	2	3	4	5
	제안 된 대책 ▪ **자원을 늘리기**											
	더 어려운 과제 할당						**새 스포츠/취미를 가지는 것**					
24	나는 더 까다로운 작업을 배정받기를 원한다.	1	2	3	4	5	나는 일을 잊어버리게 해주는 새로운 스포츠나 취미를 갖고 싶다.	1	2	3	4	5
	더 많은 휴식 시간						**주간 활동 계획**					
25	나는 근무 시간 사이에 휴식을 취하고 마음을 편안하게 할 수 있기를 원한다,.	1	2	3	4	5	친구 나 가족과 함께 주간 활동을 계획하는 것은 나에게 굉장히 기대되는 일들 이다.	1	2	3	4	5
	더 많은 자유 제공						**더 많은 개인 시간 할애**					
26	내 일에 더 많은 자유를 주면 더 나은 결과를 얻을 수 있다.	1	2	3	4	5	자유 시간에는 일과 삶에서 멀리 벗어난 더 많은 개인적인 시간을 갖고 싶다.	1	2	3	4	5
e)	**일과 삶의 갈등**											
	시간을 기반으로 한 일의 삶에서 갈등						**시간을 기반으로 한 삶의 일에서 갈등**					
27	내 일은 내가 원하는 것보다 내 삶의 활동을 방해한다.	1	2	3	4	5	삶의 책임에 사용하는 시간은 종종 내 업무적 책임을 방해한다.	1	2	3	4	5
28	직장에 헌신해야 할 때가 되면 가정의 책임과 활동에 똑같이 참여하지 못하게 된다.	1	2	3	4	5	내 삶에 보내는 시간은 종종 나의 경력에 도움이 될 수 있는 직장 생활에 시간을 투자하지 못하게 만든다.	1	2	3	4	5

29	나는 일의 책임에 소비해야 하는 시간 때문에 삶의 활동을 그리워해야 한다.	1	2	3	4	5	나는 삶의 책임에 써야 하는 시간 때문에 직장생활을 그리워해야 한다.	1	2	3	4	5
	스트레스 기반인 일의 삶에서 갈등						**스트레스 기반인 삶의 일에서 갈등**					
30	직장에서 퇴근했을 때 나는 종종 삶의 활동 / 책임에 참여하기에 너무 혼란스럽다.	1	2	3	4	5	집에서의 스트레스 때문에 직장에서 삶과 관련된 문제에 종종 몰두한다.	1	2	3	4	5
31	집에서 일할 때 감정적으로 힘들어져서 내 삶에 기여하지 못하는 경우가 많다.	1	2	3	4	5	인생에 대한 책임감 때문에 종종 압박을 받아서 일에 집중하기가 힘들다.	1	2	3	4	5
32	직장에서 받는 모든 압력 때문에 나는 집에 왔을 때 가끔 즐거운 일을 하기에 너무 압박감을 느낀다.	1	2	3	4	5	내 삶의 긴장과 불안은 종종 나의 업무 수행을 약화시킨다.	1	2	3	4	5
	행동을 기반으로 한 일의 삶에서 갈등						**행동을 기반으로 한 삶의 일에서 갈등**					
33	직장에서 하는 문제 해결 행동이 집에서의 문제를 해결하는 데 효과적이지 않다.	1	2	3	4	5	집에서 나를 위해 하는 행동들이 직장에선 효과적이지 않은 것 같다.	1	2	3	4	5
34	직장에서 나를 위해 하는 효과적이고 필요한 행동들은 집에서는 비생산적이다.	1	2	3	4	5	집에서 나를 위해 하는 효과적이고 필요한 행동들은 직장에서는 비생산적이다.	1	2	3	4	5
35	직장에서 나를 효과적으로 만드는 행동은 비 직장 생활에서 더 나은 사람이 되는 데 도움이 되지 않는다.	1	2	3	4	5	집에서 나를 위한 문제를 해결 하는 행동들은 직장에서 유용하지 않은 것처럼 보인다.	1	2	3	4	5
f)	**삶과 일의 풍요로움**											
	* 질문 : 나의 일 / 삶에 대한 나의 개입 ______________.											
	일의 삶에서의 발전						**삶의 일에서의 개발**					

36	다른 관점을 이해하는 데 도움이 되며 내 인생에서 더 나은 사람이 될 수 있다.	1	2	3	4	5	내가 지식을 얻도록 도와 주며, 이것이 내가 더 나은 직장인이 될 수 있도록 도와 준다.	1	2	3	4	5
37	지식을 얻도록 도와 주며 내 삶에서 더 나은 사람이 되도록 도와준다.	1	2	3	4	5	기술을 얻을 수 있도록 도와주며, 이것이 나를 더 나은 직장인으로 만들어 준다.	1	2	3	4	5
38	기술 습득에 도움이 되며 내 삶에서 더 나은 사람이 될 수 있다.	1	2	3	4	5	새로운 일에 대한 지식을 넓히는 데 도움이 된다. 그러면 더 나은 직장인이 될 수 있다.	1	2	3	4	5
	일의 삶에서의 영향						**삶의 일에서의 영향**					
39	좋은 분위기로 나를 끌어 들여 내 인생에서 더 나은 사람이 될 수 있게 해준다.	1	2	3	4	5	좋은 분위기로 나를 끌어 들여 더 나은 직장인 될 수 있게 도와준다.	1	2	3	4	5
40	나를 행복하게 해주고 내 인생에서 더 나은 사람이 될 수 있게 해준다.	1	2	3	4	5	나를 기쁘게 하며 이것이 내가 더 나은 직장인이 되도록 돕는다.	1	2	3	4	5
41	내가 힘이 솟게 만들며 더 나은 사람이 되도록 돕는다.	1	2	3	4	5	내가 힘이 솟게 만들며 더 나은 직장인이 되도록 돕는다.	1	2	3	4	5
	일의 삶에서의 자본						**삶의 일에서의 효율성**					
42	개인적으로 성취감을 느낄 수 있게 도와주고 내 인생에서 더 나은 사람이 될 수 있게 도와준다.	1	2	3	4	5	직장에서의 시간 낭비를 피할 수 있게 요구해서 더 나은 직장인이 될 수 있다.	1	2	3	4	5
43	성취감을 주며 이것은 내 삶에서 더 나은 사람이 될 수 있도록 도와준다.	1	2	3	4	5	집중적인 방식으로 내 업무 시간을 사용하도록 장려하는 것은 내가 더 나은 직장인이 될 수 있도록 도와 준다.	1	2	3	4	5
44	나에게 성공의 기분을 제공하고 이것은 내 인생에서 더 나은 사람이	1	2	3	4	5	나를 직장에 더 집중하게 하고 이것은 내가 더 나은 직장인이	1	2	3	4	5

	되도록 돕는다.						되도록 돕는다.					

파트 C – 일과 삶의 균형에 대한 전반적인 평가												
g)	**대처 전략**											
	동료 및 고용주의 지원						**가족 및 친구 지원**					
45	동료 및 고용주의 지원은 스트레스에 더 잘 대처하는 것과 업무 및 삶의 역할에서 나의 성과를 향상 시키는데 도움이 되고 있다.	1	2	3	4	5	친구들과 가족의 지원은 스트레스에 더 잘 대처하는 것과 일과 삶의 역할에서 나의 성과를 향상시키는데 도움이 되고 있다.	1	2	3	4	5
	레저 활동						**인지 관리**					
46	평소의 여가 활동은 스트레스에 더 잘 대처하고 일과 삶의 역할에서 나의 성과를 향상 시키는데 도움이 되고 있다.	1	2	3	4	5	나는 직장이나 삶에서의 높은 스트레스나 낮은 실적을 대처할 수 있는 능력을 갖게 하는 일상적인 일과 습관에 익숙하다.	1	2	3	4	5
	행동 관리											
47	나는 직장이나 삶에서 높은 스트레스 나 낮은 수행 능력을 다루기에 적합한 특정한 행동에 참여할 것이다.	1	2	3	4	5						

h)	**일과 삶의 균형**											
48	나는 직장에서나 내 인생에서 나에게 기대되는 것을 협상하고 성취 할 수 있다.	1	2	3	4	5	나는 상사와 나의 가족이 나에게 기대하는 것을 성취 할 수 있다.	1	2	3	4	5
49	나는 일과 삶에서 중요한 사람들의 역할에 대한 기대를 충족시키는	1	2	3	4	5	동료, 친구 및 가족은 내가 기대에 부합한다고 말할 것이다.	1	2	3	4	5

	훌륭한 일을 한다.											
50	나와 가까이 있는 사람들은 일과 삶의 균형을 잘 잡을 것이라고 말할 것이다.	1	2	3	4	5	나는 동료, 친구 및 가족의 의견을 바탕으로 내 업무와 삶의 의무를 다 수행하고 있음이 확실하다.	1	2	3	4	5

i)	**일의 영역에서 결과들**						**삶의 영역에서 결과들**					
	직업 만족						**가족 만족**					
51	직장에서 하는 일은 나에게 의미가 있다.	1	2	3	4	5	내 가족의 조건은 훌륭하다.	1	2	3	4	5
52	결국 나는 내 일에 만족한다.	1	2	3	4	5	나는 내 가족의 삶에 대하여 만족한다.	1	2	3	4	5
	회사 헌신						**인생의 만족**					
53	나는 내가 일하고 있는 사람들에게 말할 수 있다는 것을 자랑스럽게 생각한다.	1	2	3	4	5	나는 내 인생에 만족한다.	1	2	3	4	5
54	나는 내 회사의 문제가 마치 자신의 문제라고 생각한다.	1	2	3	4	5	대부분의 경우 내 삶은 내 이상에 가깝다.	1	2	3	4	5
	이직에 대한 의도						**정신의 건강**					
55	내년에 새로운 일자리를 적극적으로 찾고 있을 것 이다.	1	2	3	4	5	일반적으로 나는 자신감이 있고 나 자신에 대해 긍정적이다.	1	2	3	4	5
56	나는 이 일을 쉽게 포기하지 않을 것이다.	1	2	3	4	5	대부분의 경우, 나는 내가 누구이며, 내가 인도하는 삶에 대해 자랑스럽다.	1	2	3	4	5

□ 이 연구의 요약 보고서를 받고 싶습니다. 사본을 위의 이메일 주소로 보내주십시오.

이 설문지를 작성해 주셔서 대단히 감사합니다.

Appendix G
Verbatim Report 1

Date of interview: 2nd July 2017
Time of interview: 7 pm
Detail: A face to face interview was held on 2nd July with Ms. A, who will be referred to as "A". She has been working in a Grade A1 (General Builder) main contractor firm for one year as a Quantity Surveyor. Ms. A is currently a 24-year old Singaporean and belongs to a middle-income group family. The interviewer shall be referred to as "B". Results of the inferential statistics were discussed with Miss A prior to the start of the interview to allow her to express her opinions or provide addition explanations, if any.

<u>Interview Transcript</u>

B: Do you agree that Work Demands is important in explaining an individual's Work–Life Conflict?
A: Yes. Because it is how much your workload and your amount of responsibilities that determine how much stress you are under.
B: Do you agree that Boundary Spanning Demands is important in explaining an individual's Work–Life Conflict?
A: Less so. Usually policies that are implemented in the organization are usually with positive intention and are less of a Boundary Spanning Demand.
B: Do you agree that Life Demands is important in explaining an individual's Work–Life Conflict?
A: No. Work is still the primary consideration and life demand don't really affect much of my Work–Life Conflict. It depends if you are married maybe the perspective may change.
B: Of all the three demands, which one do you think is more likely to cause Work–Life Conflict? Why?
A: Work Demand is definitely the cause of my Work–Life Conflict. OT to rush work that is meaningful is still ok. OT like admin stuff is unpleasant. My relationship with boss is positive and the workload he gave is still manageable

L. Sui Pheng and B. K. Q. Chua, *Work–Life Balance in Construction*,
Management in the Built Environment,
https://doi.org/10.1007/978-981-13-1918-1

but sometimes when the project is demanding it really does get a little messy and eat into my life for at least a week.

B: Do you agree that Work Resources is important in explaining an individual's Work–Life Enrichment?

A: I think Work Resources is more of a job satisfaction than Work–Life Balance. It is more like learn more from work than providing Work–Life Balance to me.

B: Do you agree that Boundary Spanning Resource is important in explaining an individual's Work–Life Enrichment?

A: I think it is important but my company does not really provide such resources at the moment as you know construction projects are hard to allow team member to work from home as client may require you to be present when he needs you. If policies were to be implemented in my company that integrates work and life, this boundary resources will definitely be beneficial. But culture also plays a part. Sometimes when I OT it not because I want to but because my senior has not left her work. I feel bad because she got more work and I want to be on standby to help her.

B: Do you agree that Life Resources is important in explaining an individual's Work–Life Enrichment?

A: It will. Sometimes weekend can help let you look forward to your following weekday. Sometimes when your weekend is ruined, it may make my weekday moody. Psychological reward is a very good way in helping me feel motivated. And also when you have high work demand, sometimes spend time with friends and family can help to recuperate from the stresses.

B: Of all the three resources, which one do you think is more likely to cause Work–Life Enrichment? Why?

A: Life resources are important to Work–Life Enrichment but I want to emphasize work resources is important to job satisfaction.

B: Do you agree that reducing Work–Life Conflict and/or improving Work–Life Enrichment will lead to Work–Life Balance?

A: Yes. Both will individually contribute to Work–Life Balance.

B: It is posited that low Work–Life Conflict and high Work–Life Enrichment will result in an individual achieving Work–Life Balance. However, the results suggested otherwise. It can be seen that Work–Life Enrichment is significant towards explaining Coping Strategies but not Work–Life Conflict. What is your view on this?

A: For me I don't think it is just Work–Life Enrichment. For people who OT a lot, I feel that they require cognitive management more as they are used to it and be still able to attain Work–Life Balance. This happens to my colleagues as well because they OT a lot but they don't feel the negative effect at all and are still able to handle their non-work schedule. I feel that there is certain limit as to how much people will have in their Work–Life Conflict once it reaches the threshold, I don't think Work–Life Enrichment will be beneficial to explain Coping Strategies. Reduce the WLC then you can think of the WLE. Maybe your survey of all the respondents have not reach the threshold that is why WLE is significant.

B: Do low Work–Life Conflict and high Work–Life Enrichment lead to Work–Life Balance? Or vice versa?
A: I think for blue collar, low WLC is enough for them to attain WLB as they don't expect too much from their job. White collar then we must factor in high WLE. For me I definitely feel that both contribute to higher level of Work–Life Balance.
B: Do you agree that even if conflict or enrichment is the likely consequence of engaging in work and family roles, it is how people cope with conflicts that determines their personal outcomes?
A: Yes Coping Strategies definitely determine my personal outcome especially the mentality management (cognitive). Everyone can feel that they have no WLB but if he or she thinks is acceptable then is the mentality play a role in helping that person attain WLB.
B: Do you think that Work–Life Enrichment has a stronger effect in explaining Coping Strategies than Work–Life Conflict?
A: I think it depends. I feel that if you have high WLC, it is the matter of your cognitive and behavioral management that helps you to cope with WLC in order to attain WLB. But for WLE, I feel that it is the support and leisure activities that help you improve your WLE. But that doesn't mean the rest are not important; it is just it may not contribute substantially.
B: Do you agree that coping strategies will alter your perception on Work–Life Balance whereby positive coping strategies will lead to the achievement of Work–Life Balance?
A: Yes I agree. Especially the expectation in my job if I have a high expectation in my work, coping strategy will be important in helping me cope with the negative effect in order for me to get my job done. If I want to climb the corporate ladder, coping strategy must be with me all the time.
B: Do you agree that achieving Work–Life Balance will lead to job satisfaction, organizational commitment, reduce in turnover intention, family satisfaction, life satisfaction and psychological well-being?
A: Yes but if I weigh them, I feel that WLB will lead to job, life and family satisfaction but not on organizational commitment and reduce in turnover intention. For organizational commitment, it is more on people related factor like relationships with colleagues and bosses that lead to your commitment. Turnover intention is more on opportunity if given an opportunity for higher salary and career advancement I will quit my job immediately but because of Work–Life Balance I stay in my job is no for me. Psychological well-being I don't feel it now, it does not apply to me now.

Appendix H
Verbatim Report 2

Date of interview: 12th July 2017
Time of interview: 10 am
Detail: A phone interview was held on 12th July with Manager B, who will be referred to as "Mr. B". He works in a Grade A1 (General Builder) main contractor firm for eleven years as a Mechanical and Electrical Contracts Manager. Mr. B holds the management position and currently has six Millennials employees working under him. The interview was conducted over the phone as requested by Mr. B. Results of the inferential statistics were sent to Mr. B via email prior to the start of the interview to allow him to understand the conceptual framework.

Interview Transcript

B: Do you agree that Work Demands is important in explaining an individual's Work–Life Conflict?
Mr. B: Yes. I think that everyone will have work demand is just the degree of impact it has on the individual.
B: Do you agree that Boundary Spanning Demands is important in explaining an individual's Work–Life Conflict?
Mr. B: Yes. I think it will have impact especially the amount of OT we have.
B: Do you agree that Life Demands is important in explaining an individual's Work–Life Conflict?
Mr. B: Yes. Same as work demand is the matter of degree of impact on the individual. Different people will have different impact at certain times.
B: Of all the three demands, which one do you think is more likely to cause Work–Life Conflict? Why?
Mr. B: I think work demand will have the major impact because our work is very demanding because the work volume can be a lot at times and it can affect their life as well. One example will be one of my staff wants to apply three weeks leave but I only can grant her two weeks leave due to the amount of work

L. Sui Pheng and B. K. Q. Chua, *Work–Life Balance in Construction*,
Management in the Built Environment,
https://doi.org/10.1007/978-981-13-1918-1

load she has. I understand it may cause dissatisfaction but no choice as three weeks is too long.

B: Do you agree that Work Resources is important in explaining an individual's Work–Life Enrichment?

Mr. B: Yes. Work resource is definitely needed. Interpersonal skills are one of the things that are applicable to life use which is brought over from work. Training resources provided in the company also helps the employee upgrade their skills and knowledge and benefit them to progress further.

B: Do you agree that Boundary Spanning Resources is important in explaining an individual's Work–Life Enrichment?

Mr. B: Yes it does. Our company will try to provide such resources annually like for example family day or dinner and dance event. During the last family day, we went to the zoo and all the employees will bring their loved ones along. This is a good way to allow the employee's family to interact with their colleagues or bosses.

B: Do you agree that Life Resources is important in explaining an individual's Work–Life Enrichment?

Mr. B: Yes it will. Foreign talents in Singapore may be less likely as they do not have their loved ones here. For Singaporeans, it will definitely as the life resource can help to compensate the demand that the employee currently has.

B: Of all the three resources, which one do you think is more likely to cause Work–Life Enrichment? Why?

Mr. B: I think Work and Life resources are important for the Singaporeans as these are the very important way to help reduce their demand. Boundary spanning resource less likely as the number of times this situation happen in our company is quite less (once a year).

B: Do you agree that reducing Work–Life Conflict and/or improving Work–Life Enrichment will lead to Work–Life Balance?

Mr. B: Both will have an impact. Conflict will be one major issue but with enrichment in the picture, I think that the Millennials are able to overcome the conflict even if there is high conflict.

B: Do you agree that even if conflict or enrichment is the likely consequence of engaging in work and family roles, it is how people cope with conflicts that determines their personal outcomes.

Mr. B: Yes I agree. I find that work life conflict cannot be chunked aside as I felt that it will have an adverse effect on an individual work life balance. That is why you need coping strategies to cope with the conflict. Enrichment must always be available to the individual and we must try to provide enhancement strategies to help them increase their level of enrichment.

B: It is posited that low Work–Life Conflict and high Work–Life Enrichment will result in an individual achieving Work–Life Balance. However, the results suggested otherwise. It can be seen that Work–Life Enrichment is significant towards explaining Coping Strategies but not Work–Life Conflict. What is your view on this?

Mr. B: I find that conflict and enrichment will definitely affect work life balance but as you introduce coping strategies as intermediate strategy, it helps to nullify the conflict and maximize the enrichment that helps the worker achieve work life balance. And also, I felt that even though workers may encounter major conflicts in their work, if enrichment is present it can help to overshadow the high conflict which is why conflict is not associated to work life balance in your study. I agreed with your explanation and will like to add on. Reducing conflict itself is not enough in my view but with the enrichment available, the Millennials can be more likely to achieve work life balance. Maybe it is because in the construction industry there may not be a lot of enrichment available which is why coping strategies is needed to overcome the low enrichment.

B: What is your suggestion in helping Millennials employees attain Work–Life Balance? (Share your experiences).

Mr. B: Sometimes I will organize a chit-chat session with my staffs to talk about their work and life matters. By understanding my staffs outside work matters, I will know how to deal with him/her better.

B: Do you agree that coping strategies will alter your perception on Work–Life Balance whereby positive coping strategies will lead to the achievement of Work–Life Balance?

Mr. B: Yes I agree. Coping strategies must also be present to help contribute to work life balance of an individual.

B: Do you agree that achieving Work–Life Balance will lead to job satisfaction, organizational commitment, reduce in turnover intention, family satisfaction, life satisfaction and psychological well-being?

Mr. B: Yes isn't it obvious? But for turnover intention, it depends on the pull factor. As the job here is very demanding and the pay is below or average market rate, given the opportunities (good pay or more career advancement) they will leave the company no doubt. Work life balance itself is not enough to retain them.

Appendix I
Verbatim Report 3

Date of interview: 21st August 2017
Time of interview: 7 pm
Detail: A face-to-face interview was held on 21st August with Mr. C, who will be referred to as "C". Mr. C is a 36-year old South Korean who has been working in Singapore for two years. He was deployed to Singapore by his company and is now working in a project located in the east part of Singapore. He has been working with the South Korean main contractor firm for ten years and is currently holding the senior site engineer position. Results of the inferential statistics were discussed with Mr. C prior to the start of the interview to allow him to present his views or to add any further points, if any.

Interview Transcript

B: Do you agree that Work Demands is important in explaining an individual's Work–Life Conflict?
C: I have already worked in this company for ten years and my position is getting higher. My job requirements also increase. My working scope has increased and I have to take care of more people and activities. That means I cannot be free from work even when I go back home. They don't care whether I end work at 2 am or 3 am because it is my duty to take care of the overall project. The more work I receive, the more stressed I am. This is where the conflict part comes in.
B: Do you agree that Boundary Spanning Demands is important in explaining an individual's Work–Life Conflict?
C: Yes also. Just like I explained to you, even I go back home without OT, without OT payment I still have to work. Sometimes I have work over the holiday and Saturday or even some Sundays without any OT payment. Our company does not give us OT payment. Even I go back home I cannot be free from work.
B: Do you agree that Life Demands is important in explaining an individual's Work–Life Conflict?

L. Sui Pheng and B. K. Q. Chua, *Work–Life Balance in Construction*,
Management in the Built Environment,
https://doi.org/10.1007/978-981-13-1918-1

C: Yes of course. I am a father and I am a husband. I have the duty to support the family. I also have the duty for this project. My wife will complain why I still work even when I come back home, I usually explain to her but she sometime don't understand. Why I work here is to support them and make them happy. Of course I should balance work and family but I spent too much time on work causing conflict. I cannot satisfied both. I only have 24 h.

B: Of all the three demands, which one do you think is more likely to cause Work–Life Conflict? Why?

C: Of course Work Demand. Similar to what I have explained just now.

B: Do you agree that Work Resources is important in explaining an individual's Work–Life Enrichment?

C: Promotion and increasing salary are the reasons I came here and this one of course if I can earn more money I can feel motivated. Money is still important for me. Sometimes I also feel achievement when I settled a challenging work.

B: Do you agree that Boundary Spanning Resource is important in explaining an individual's Work–Life Enrichment?

C: No. I currently don't have flexible work schedule or work at home. As you know, South Korean culture is very rigid not flexible. Our hierarchy is also very strong and our working hours are from 7 am to 6 pm. Construction team had to work 24 h around the clock because we have to think about the actual site conditions. Have to monitor the site.

B: Do you agree that Life Resources is important in explaining an individual's Work–Life Enrichment?

C: Actually just six months ago I brought my family here. Before that we lived separately. When we live together, I feel very happy. When I go home to play with my daughter even though 10 min or 15 min I can smile. I work for the sake of my family. Sometime I can feel stress but when I see my daughter I will feel happy.

B: Of all the three resources, which one do you think is more likely to cause Work–Life Enrichment? Why?

C: Life resources. Because I am working here because of my family.

B: Do you agree that reducing Work–Life Conflict and/or improving Work–Life Enrichment will lead to Work–Life Balance?

C: Both are very important. Of course WLC should decrease and WLE should increase and they can meet at one point which is the WLB.

B: It is posited that low Work–Life Conflict and high Work–Life Enrichment will result in an individual achieving Work–Life Balance. However, the results showed that only Work–Life Enrichment is significant in contributing to Work–Life Enrichment but not Work–Life Conflict. What is your view on this?

C: Actually for me Work–Life Enrichment for us is increase in salary and make more money. In Singapore and Korea I think money is very important for them and us to survive. South Korea's economy is not good, very hard to find job. They will compare with their friends or other persons who earn more money. WLC we can control one time we adjust it, it can become normal. If they earn more money they can bear more stress.

B: The results showed that Work–Life Enrichment is significant in explaining Coping Strategies but Coping Strategies does not contribute to Work–Life Balance. What is your view on this?
C: Having coping strategies is good but it is only temporary. Maybe you go overseas relax they still need to work. For example, before my family came here I will have traveled back to Korea every four months to visit them. This regular holiday can be a coping strategy. When I go back Korea I can relax without work pressure but after one week I will start concerning about my work. Because when I go back to work everyone will wait for me.
B: Do you agree that even if conflict or enrichment is the likely consequence of engaging in work and family roles, it is how people cope with conflicts that determines their personal outcomes.
C: I have no comment for this statement.
B: Do you agree that coping strategies will alter your perception on Work–Life Balance whereby positive coping strategies will lead to the achievement of Work–Life Balance?
C: Of course it will help even though I find that it is temporary. Even though your result indicates that it is not significant may be is because it is less effective.
B: Do you agree that achieving Work–Life Balance will lead to job satisfaction, organizational commitment, reduce in turnover intention, family satisfaction, life satisfaction and psychological well-being?
C: This one is very important as I mentioned I am married. Work–Life Balance will definitely help in my family satisfaction. Work also if I have WLB I will have less stress which help me be more effective in my work and make me become more happy.
B: How is the working environment in Singapore as compared to South Korea? What are the differences between your Singaporean colleagues and South Korean colleagues?
C: In South Korea, all my colleagues are South Koreans and I can talk to them using my language. But in Singapore I have to use English but not all of them understand English making me very hard to communicate with them especially the foreigners. There are so many foreign workers here and only a few Singapore people. My project here (Singapore) always need me to work overtime until very late. When I reached home, it is already 9 pm. All my Singaporean colleagues worked until very late also. In South Korea, I will knock off at 6 pm. One thing I don't like about this project (in Singapore) is the site is very messy and dirty unlike Korea things are properly placed.

Appendix J
Verbatim Report 4

Date of interview: 22nd August 2017
Time of interview: 11 am
Detail: A face-to-face interview was held on 22nd August with Director D, who will be referred to as "D". He has been working in a Grade A1 South Korean main contractor firm for more than six years and is currently a project director for a project in the east part of Singapore. Director D is currently 47-year old and he has worked with several South Koreans Millennials employees who are either full-time or deployed to Singapore for on-the-job training. Inferential statistics on the South Korean results were discussed with Director D prior to the start of the interview to allow him to present his opinions or to add further points, if any.

Interview Transcript

B: Do you agree that Work Demands is important in explaining an individual's Work–Life Conflict?
D: Well of course it's important. Work is always the main source of problem regardless how great your life is. Work demand causes work and life conflicting with each other no doubt.
B: Do you agree that Boundary Spanning Demands is important in explaining an individual's Work–Life Conflict?
D: Definitely. You cannot avoid OT especially during peak period where work has to be done as soon as possible. Bringing work back home is very common for us not just for the South Koreans.
B: Do you agree that Life Demands is important in explaining an individual's Work–Life Conflict?
D: I don't think it will affect that much for the South Korean staffs. They do not have friends or families here so I think life demand may not be so impactful towards the work life conflict.

L. Sui Pheng and B. K. Q. Chua, *Work–Life Balance in Construction*, Management in the Built Environment,
https://doi.org/10.1007/978-981-13-1918-1

B: Of all the three demands, which one do you think is more likely to cause Work–Life Conflict? Why?

D: Work demand, construction is not an easy job even for people like me. I felt that the South Koreans in Singapore here have a hard time communicating with us or even the foreigners which make their task two times harder to complete. It takes time for them to adapt and blend into our culture. It is tough for them at first maybe it is because South Korea's construction is different from Singapore's construction and their work ethics have to change to suit our requirements.

B: Do you agree that Work Resources is important in explaining an individual's Work–Life Enrichment?

D: I think this is quite important. It is like a source of motivation for them. The South Koreans I met are all very hardworking and during my conversation with one of the staffs over lunch break, she told me that stationing in Singapore can help them progress faster.

B: Do you agree that Boundary Spanning Resources is important in explaining an individual's Work–Life Enrichment?

D: Not quite. Enticing the staffs by giving them time off to do their other stuff I don't think can help enrich their personal life. The thing you said like annual dinner and dance we do have but once a year I don't think can help them much. If you say flexible work arrangement, yes it will help but I don't allow it because you may never know when I need them.

B: Do you agree that Life Resources is important in explaining an individual's Work–Life Enrichment?

D: Of course it is important but not that much for them because they don't have friends and families here. Assume that they are working in South Korea, then I would say that life resources is definitely important in enriching their work and life.

B: Of all the three resources, which one do you think is more likely to cause Work–Life Enrichment? Why?

D: Two scenarios here. Work Resources is when South Koreans are in Singapore because deploying to Singapore maybe beneficial towards their career advancement and they are enticed by it. Life Resources is when they are in their home country where friends and families are around they will definitely need these resources to overcome the work demand.

B: Do you agree that reducing Work–Life Conflict and/or improving Work–Life Enrichment will lead to Work–Life Balance?

D: Yes it does given that reduce negative and increase positive is always beneficial to anyone.

B: Do you agree that even if conflict or enrichment is the likely consequence of engaging in work and family roles, it is how people cope with conflicts that determines their personal outcomes.

D: Definitely. If you every day face with high demand causing work life conflict you cannot expect the conflict to stay with you throughout the week. Like I said earlier, the South Koreans may face issues adapting to our culture but as times goes by they started to blend in. Same logic.

B: It is posited that low Work–Life Conflict and high Work–Life Enrichment will result in an individual achieving Work–Life Balance. However, the results showed that only Work–Life Enrichment is significant in contributing to Work–Life Enrichment but not Work–Life Conflict. What is your view on this?

D: Maybe it is due to the fact that the South Koreans are more hardworking and think that they will be rewarded if they work hard. High enrichment as a form of career progression for them is desirable especially for those who are willing to sacrifice their time with friends and families and work in Singapore. Low conflict is not significant maybe is because the South Koreans have already knew that in construction you will definitely face with conflicts of interest between work and life. They choose to work in construction means they knew they will face these consequences and willing to make this sacrifice.

B: The results showed that Work–Life Enrichment is significant in explaining Coping Strategies but Coping Strategies does not contribute to Work–Life Balance. What is your view on this?

D: I don't totally agree on this question. Coping Strategies is needed to meet your Work–Life Balance. You cannot expect a guy without any coping strategy to answer you that he has work life balance. Everyone will have their own coping strategy but maybe different from each other.

B: Do you agree that coping strategies will alter your perception on Work–Life Balance whereby positive coping strategies will lead to the achievement of Work–Life Balance?

D: Yes like I said earlier you need coping strategies whenever you feel that you are not doing well in life.

B: Do you agree that achieving Work–Life Balance will lead to job satisfaction, organizational commitment, reduce in turnover intention, family satisfaction, life satisfaction and psychological well-being?

D: Yes but maybe not family satisfaction because they are stationed in Singapore.

Lightning Source UK Ltd.
Milton Keynes UK
UKHW02n1328280818
327921UK00001B/13/P

9 789811 319174